An Introduction
to Animal Law

An Introduction to Animal Law

Margaret E. Cooper, LLB

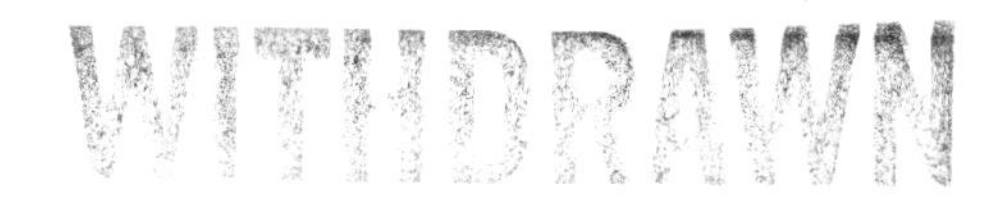

1987

ACADEMIC PRESS

Harcourt Brace Jovanovich, Publishers
London San Diego New York
Berkeley Boston Sydney Tokyo Toronto

ACADEMIC PRESS LIMITED
24–28 Oval Road
London NW1

United States Edition published by
ACADEMIC PRESS INC.
Orlando, Florida 32887

This publication is designed to provide useful information in regard to the subject matter covered. It is sold with the understanding that the publisher is not engaged in rendering legal services or providing legal advice concerning particular matters

British Library Cataloguing in Publication Data
Cooper, Margaret E.
An introduction to animal law
1. Animals, Treatment of—Law and legislation
I. Title
342.647 K3620

ISBN 0-12-188030-3

LCCN 87-72116

Typeset by Colset Private Limited, Singapore
Printed by St Edmundsbury Press
Bury St Edmunds, England

TO

my Father

FRANCIS H. VOWLES, LLB, Solicitor
who trained me in his profession
and fostered my interest in animal law

We must not make a scarecrow of the law,
Setting it up to fear the birds of prey,
And let it keep one shape, till custom
make it
·*Their perch, and not their terror.*

("Measure for Measure" II, i, 1., William Shakespeare)

Foreword

As Margaret Cooper herself acknowledges in her opening chapter, the law relating to animals has so many facets, and can be approached from so many angles, that it is virtually impossible to cover the subject comprehensively. Sensibly, therefore, she has chosen to write a practical book which will provide helpful guidelines for those who wish to be pointed in the right direction but not to be overwhelmed by detail.

The aspects of the law which she has chosen to cover and her method of presentation provide evidence of her ability to marry her knowledge of the law to an awareness of animals as creatures with genuine rights of their own — a reflection perhaps of her own background as a lawyer married to a veterinary surgeon.

It is a privilege to welcome this addition to the regrettably meagre collection of books on animal law and I commend it to all who are involved with animals in any way.

ALASTAIR PORTER

Preface

Legislation relating to animals has ancient origins and in many civilizations certain species have held particular significance, be it religious, cultural, nutritional or sporting; some of the earliest laws on this subject were carved in stone — the hieroglyphics of the 18th century BC *Codex of Hammurabi* proclaimed it illegal to overwork animals (Vogel-Etienne, G. (1980) *Animals International* **1** (2), 12).

As a general rule, however, throughout the world the law was primarily concerned with animals as property, rather than in need of protection, until the 19th century. Since the 1970s animal law has proved to be a growth area in the production and enforcement of both national and international legislation. This has been particularly so in the areas of conservation and welfare and there has been extensive legal and philosophical consideration of the status of animals.

As Chapter 1 makes clear, this book is not intended to be a standard text, but a guide for the lay person — namely, to help the non-lawyer to understand the basic concepts of animal law and to provide the lawyer (who is a lay person in the world of animal science) with an introduction to relevant concepts and literature which are not normally found in the conventional legal texts.

I should like to thank all those, particularly from the fields of veterinary and biological science, who have encouraged me in my study of animal law and its related topics by providing me with information and by persuading me to write, lecture or simply to answer their questions or to peruse their own writings. I should like to acknowledge in particular Dr C.R. Coid who suggested that I write this book, the long-standing help of many of those mentioned later, together with Messrs A.G. Greenwood and G. Joss for raising legal issues, the late Dr D.I. Chapman and Mrs Norma Chapman for introducing me to the world of deer and for their valued friendship, Dr H. Rozamund, Dr J. Shuja and Mr Brian Vernon for giving me books about the Netherlands, Mauritian and British veterinary services respectively, Dr H. Rowsell and many others for literature from their respective countries, members of the Conservation Monitoring Centre and the International Council for Bird Preservation at Cambridge for the exchange of international legislation, Mr J.T. Eley for providing literature and for practical discussions on conservation law over the years and members of the Headquarters staff of the RSPCA who provided me with stop-press news of legislation which was developing as I finished this book. The World Society for the Protection of Animals has given me most welcome support and encouragement in my work. The libraries of the Royal College of Veterinary Surgeons and the Royal College of Surgeons of England have always been most helpful and the Squire Law Library of Cambridge University and the Law Society's Library have been valuable sources of information.

I especially appreciate the comments and advice of Messrs J.E. Cooper, A.M. Taylor and G.S. Wiggins, Dr A. Knifton, Mr A.R.W. Porter, Mr F.H. Vowles, Mr C. Platt of the World Society for the Protection of Animals, Mr P.N. O'Donoghue of the Institute of Biology, Dr L. Batten and Mr D. Morgan of the Nature Conservancy Council and Dr M. Buttolph of the University of London, who read draft versions of various chapters of this book.

Collecting apposite quotations has become something of a hobby and has often provided a sense of achievement as much from the finding as from the intrinsic suitability of the words themselves. I must admit to several "second-hand" discoveries and acknowledge those who, although the originals may be there for all to use, brought them to light, namely the late Emily Stuart Leavitt of the Animal Welfare Institute (see Chapter 3), Dr D. Clarke (in *Antenna* **5** (2), 67) (see Chapter 3) and Dr J. Shuja of the Mauritius Society for the Prevention of Cruelty to Animals (see Chapter 9).

Miss S. Dowsett and Mrs S. Milne have kindly typed manuscripts for me and special thanks must go to Mrs D.M. Suttie who, for nine years and with enduring patience, has produced faultless typescripts in the face of handwriting and output which could, at best, be described as erratic.

I should like to thank my family for all their help and support. My parents, Francis and Elizabeth Vowles, have always taken an interest in my work and supplied relevant data; my parents-in-law, Eric and Dorothy Cooper, have provided encouragement, companionship, tea and sympathy over the years.

My daughter, Vanessa, and my son, Maxwell, have foregone home comforts and helped with proofreading, indexing and domestic arrangements on the understanding that they might, one day, have at least one decent meal to celebrate the publication of this book.

Finally, but otherwise first and foremost, it was my husband, John E. Cooper, FRCVS, who introduced me to the subject of animal law when we were students at the University of Bristol in the early 1960s, giving me an interest to pursue both at home and abroad ever since. He has provided the vital link with his profession as well as others and is my chief source of information on the law and its practical application. His drive and enthusiasm have enabled me to write this book, which owes much to his textual and editorial skills. John, *asante sana kwa msaada yako*.

MARGARET E. COOPER
18th September 1986

Table of Statutes: Great Britain

Note: In some parts of the book only the main Acts have been mentioned. The reference books listed in Chapter 1 should be consulted to find amending or subsequent legislation.

Table of Statutory Instruments

Note: In some parts of the book only the main Statutory Instruments have been mentioned. The reference books listed in Chapter 1 should be consulted to find amending or subsequent legislation.

Table of Other Legislation

Note: Only the main instruments of legislation have been given. There may be subsequent or amending laws.

National legislation

Statute titles are given in English

European legislation

International legislation

Table of Cases

Air India *v.* Wiggins [1980] 2 All E.R. 593, HL(E), 36
Auer (Vincent) *v.* Ministère Publique (No. 2)(No. 271/82) [1985] 1 C.M.L.R. 128, European Court, 179
Barnard *v.* Evans [1925] 2 K.B. 794, 28
Behrens *v.* Bertram Mills Circus Ltd. [1957] 1 All E.R. 583, 18
Blades *v.* Higgs (1865) 11 H.L. Cas. 621, 15
British Airways Board *v.* Wiggins [1977] 3 All E.R. 1068, 36
Chalmers *v.* Diwell [1976] Crim. L.R. 134, D.C., 38
Filburn *v.* People's Palace & Aquarium Co. Ltd. (1890) 25 Q.B.D. 258, 14
Ford *v.* Wiley (1889) 23 Q.B.D. 203, 28
Hudnott *v.* Campbell (1986) *The Times*, 27 June 1986, 28
Kavanagh *v.* Stokes [1942] I.R. 596, 19
Kirkland *v.* Robinson (1986) *The Times*, 4 December 1986, 123
Leakey *v.* National Trust [1980] 1 All E.R. 17, 17
McEwan *v.* Roddick [1952] N.Z.L.R. 938, 28
McQuaker *v.* Goddard [1940] 1 All E.R. 471, 14, 19
Murphy *v.* Zoological Society of London [1962] C.L.Y. 68, 19
Rienks (H.G.) (No. 5/83) [1985] 1 C.M.L.R. 144, European Court, 179
Rowley *v.* Murphy [1964] 1 All E.R. 50, 28, 113
RSPCA *v.* Woodhouse (1984) C.L. 693, 11
Starling *v.* Brooks [1956] Crim.L.R. 480, 31
Steele *v.* Rogers (1912) 76 J.P. 150, 14, 28
Swan *v.* Saunders (1881) 45 J.P. 522, 28
Swans, Case of (R *v.* Lady Joan Young) (1592) 7 Co. Rep. 15b., 14

Abbreviations

All E.R.	All England Law Reports
C.M.L.R.	Common Market Law Reports
K.B.	King's Bench
H.L. Cas.	House of Lords Cases
Crim.L.R.	Criminal Law Review
Q.B.D.	Queen's Bench Division
I.R.	Irish Reports
N.Z.L.R.	New Zealand Law Reports
C.L.Y.	Current Law Yearbook
C.L.	Current Law
J.P.	Justice of the Peace Reports

Table of Abbreviations

AAALAC	American Association for Accreditation of Laboratory Animal Care
ADAS	Agricultural Development and Advisory Service
APHIS	Animal Plant and Health Service
ASRA	Association for the Study of Reptiles and Amphibians
BASC	British Association for Shooting and Conservation
BFSS	British Field Sports Society
BKSTS	British Kinematograph Sound and Television Society
BVA	British Veterinary Association
CAHPA	Ad Hoc Committee of Experts for the Protection of Animals
CCAC	Canadian Council on Animal Care
CIOMS	Council for International Organizations of Medical Sciences
CITES	Convention on International Trade in Endangered Species of Wild Fauna and Flora
CoEnCo	Council for Environmental Conservation
DHSS	Department of Health and Social Security
DOE	Department of the Environment
EEC	European Economic Community
ESA	Endangered Species (Import and Export) Act 1976
FAWC	Farm Animal Welfare Council
FBCN	Fundação Brasileira para a Conservação de Natureza
FDA	Food and Drug Administration
FVE	Federation of Veterinarians in Europe
GLP	Good Laboratory Practice
HMSO	Her Majesty's Stationery Office
HO	Home Office
HS	Home Secretary (Secretary of State for the Home Office)
HSC	Health and Safety Commission
HSE	Health and Safety Executive
HSWA	Health and Safety at Work etc. Act 1974
IASP	International Association for the Study of Pain
IAT	Institute of Animal Technology
IATA	International Air Transport Association
ICLAS	International Council for Laboratory Animal Science
IOB	Institute of Biology
IUCN	International Union for Conservation of Nature and Natural Resources
JACOPIS	Joint Advisory Committee on Pets in Society
JCCBI	Joint Committee for the Conservation of British Insects
LAEG	Liverpool Animal Ethical Group
LRK	Licensed rehabilitation keeper
MAFF	Ministry of Agriculture, Fisheries and Food. The abbreviation also takes in the Agriculture Department of the Welsh Office and the Department of Agriculture and Fisheries for Scotland (see Chapter 1)
NCA	National Council for Aviculture
NCC	Nature Conservancy Council
NIH	National Institutes of Health of the Public Health Service
OGL	Open general licence

P	Pharmacy medicine
PML	Merchants', or Farmers', List medicine
POM	Prescription only medicine
RCVS	Royal College of Veterinary Surgeons
RDS	Research Defence Society
RP	Registered premises
RS	Royal Society
RSPB	Royal Society for the Protection of Birds
RSPCA	Royal Society for the Prevention of Cruelty to Animals
s.	Section (in Acts of Parliament)
ss.	Sections (in Acts of Parliament)
Sched.	Schedule (in Acts of Parliament)
SLC	Scottish Law Commission
UFAW	Universities Federation for Animal Welfare
UK	United Kingdom
USA (US)	United States of America
USDA	United States Department of Agriculture
VAT	Value added tax
WCA	Wildlife and Countryside Act 1981
WTMU	Wildlife Trade Monitoring Unit

Contents

1 Animal Law: An Introduction

This is the law of the jungle — as old and as true as the sky,
And the wolf that doth keep it shall prosper but the wolf that
doth break it must die.

("The Law of the Jungle", Rudyard Kipling)

The fox was Clerk and notar in the cause,
The Gled — the Graep (hawk and vulture) at the bar could stand.
As Advocates expert into the laws
The Dogis plea together took on hand,
Who were confederate straitly in a band
Against the sheep to procure the sentence,
Though it was false — they had no conscience.

(The Tale of the Dog, the Sheep and the Wolf.
A Legal Fable. In "Lectures on Scotch Legal
Antiquities", Cosmo Innes, 1872)

The law relating to animals is a subject of well nigh infinite dimensions. Where does one draw the line? With quadrupeds as does Sandys-Winsch (1984a)? With civil liability like North (1972) and Williams (1939)? With veterinary law (RCVS, 1987)? Or with an individual species (Gregory, 1974; Weatherill, 1979; Sandys-Winsch, 1984b; Cassell, 1987) or one area of law (Thomas, 1975)?

The dilemma of the author, particularly in the case of animal law, is that readers will undoubtedly complain that certain areas, important to themselves, are excluded and that other topics are quite irrelevant to the main purpose of the book.

This book has been compiled from the information that the author has collected in the course of speaking and writing on the subject of animals and the law, largely for those in the veterinary and biological professions. The choice of subject matter has, in a way, been directed by such people in that it has been prepared in response to their requests for lectures and articles. As such, the material has been selected for its relevance to such a readership.

It is also hoped that the book will be of value to lawyers. Animal law, particularly in matters of welfare, is increasingly attracting litigation and it is important that solicitors and barristers should be able to obtain the background information on the subject which is not normally available in existing legal literature and also that they should look at the law from the point of view of those who work with animals.

The book does not aspire to academic heights nor attempt to cover all the known cases or material on a subject which could be obtained from reading the more detailed texts. It does not offer legal advice. It is intended to present a general description of the relevant law in a manner comprehensible to the non-lawyer and to provide as wide a reference to other appropriate material as possible.

For this reason, it is important to include a caveat that no-one should rely on this book as the ultimate authority on any point of law. While every effort has been made to state matters accurately, simplification and generalisation, intended to make the subject more palatable to the reader, inevitably lead to a less-than-perfect statement of the law. In addition, case law and legislation both change from time to time. The animal health field is particularly prone to alteration. Consequently, the reader should satisfy himself that he has access to the most recent version of any legislation or case law. In particular, it should be mentioned that in this book the main Act or order is given for simplicity; subsequent amendments, unless particularly significant, have been omitted.

Any person who has a problem involving a point of law should seek advice from a solicitor in private practice. Those for whom such professional assistance is too expensive should avail themselves of the legal advice schemes or fixed-fee interviews offered by many solicitors or should get initial help from a legal advice centre or a Citizens' Advice Bureau.

This chapter will deal with a number of topics which are of general application throughout the book in order to reduce the incidence of repetition in the rest of the text.

DEFINITIONS

Gender

The masculine gender is used in the book since section 6 of the Interpretation Act 1978 provides that references to the masculine gender include the feminine (and *vice versa*). This has been followed in the text to avoid the cumbersome style involved in referring to both sexes although there is no question that, as to both humans and (except where the context makes it obvious) animals, the law is applicable alike to either gender.

The Act also provides that in legislation the term "person" includes bodies corporate, such as companies and institutions founded on Royal Charter or Acts of Parliament, as well as individuals. For animals, see Appendix 3 (Note 2).

The Crown is not bound by a statute unless it specifically so provides; nevertheless ministerial departments, whose authority is derived from the Crown, normally comply voluntarily with legislation.

Territorial Application of Law

This book deals primarily with the law applying to England and Wales. Much of it extends also to Scotland but it must be borne in mind that civil law often differs there and the judicial system is not the same. Acts of Parliament do not always apply to Scotland; there may be separate legislation, as in the case of the Protection of Animals Acts 1911–1964 and the Protection of Animals (Scotland) Act 1912 in which the initial Acts of 1911 and 1912 are separate in application but subsequent legislation applies to both Acts. The Veterinary Surgeons Act 1966, Animals (Scientific Procedures) Act 1986 and Endangered Species (Import and Export) Act 1976 apply to Northern Ireland in addition but in the main part that country has separate legislation. This has been listed for many areas of animal and veterinary law in RCVS (1987). The Channel Isles and the Isle of Man, which are part of the "British Islands" (Interpretation Act 1978, Schedule 1), more commonly referred to as the "British Isles", have separate, but often comparable, legislation and legal systems. Legislation often refers to Great Britain or the United Kingdom. These terms comprise:

Great Britain	England, Wales and Scotland
United Kingdom	Great Britain and Northern Ireland

In Chapter 9 international legislation is discussed and that of countries other than Great Britain on the topics covered by the preceding chapters.

Animal

The word "animal" must be a good contender for having the widest variety of meanings in English law. As far as possible the appropriate legal definitions have been given in the text as this is a vital element in understanding the application of each field of law. The widest statutory definition so far is to be found in the Zoo Licensing Act 1981:

> animals of the classes Mammalia, Aves, Reptilia, Amphibia, Pisces and Insecta and any other multi cellular organism that is not a plant or a fungus . . .

Another very broad definition occurs in the Transit of Animals (General) Order 1973, made under the Animal Health Act 1981 (see Chapter 3). It may come as no surprise, therefore, to find that the veterinary literature describes two species not known to science (Blackmore, 1972; Cooper and Cooper, 1982) including in the latter case the legal status.

Lawyers are not overly conscious of the discipline of taxonomy and tend to use words such as ''animal'' and ''species'' very loosely when not referring to a specific definition. In legislation and elsewhere birds are treated as distinct from animals: the words ''beast'' and ''creature'' occur occasionally. Thus, the Animal Health Act 1981 defines animals as:

> cattle [bulls, cows, steers, heifers and calves]
> sheep and goats and
> all other ruminating animals and swine

This may be expanded, by statutory instrument, to include

> any kind of mammal except man; and any kind of four-footed beast which is not a mammal

and

> fish, reptiles, crustaceans, or
> other cold-blooded creatures of any species

The Act then defines ''poultry'' as

> domestic fowls, turkeys, geese, ducks, guinea-fowls and pigeons, and pheasants and partridges

It also provides that ''poultry'' may, by statutory instrument, be extended to ''comprise any other species of bird''. This approach has its origin in the fact that animal health legislation was first instigated to control disease in farm species and has gradually been extended to cover others as the need arose in new health and welfare provisions.

The use of scientific (Latin and/or Greek) names to identify animals has only recently been a feature of legislation and then usually in connection with wild or non-indigenous species as in the Wildlife and Countryside Act 1981, other close season legislation, the Endangered Species (Import and Export) Act 1976 (see Chapter 7) and the Dangerous Wild Animals Act 1976 (see Chapter 3). This is not the practice in the animal health and animal welfare legislation (see Chapters 3–5). The scientific names of animals mentioned in this book are listed in Appendix 2.

Living things are divided into the animal and plant kingdoms, the animal kingdom being categorised as in Table 1 and Fig. 1. However, Parliament does not have to accept this approach and may lay down whatever definition it chooses. It is an essential aspect of parliamentary sovereignty that it is only constrained by the bounds of practicality from legislating that male is female and *vice versa*.

TABLE 1 Examples of Classification of Animals

	DOG	*PEREGRINE*	*HONEY BEE*
Kingdom	Animalia	Animalia	Animalia
Phylum	Chordata	Chordata	Arthropoda
Class	Mammalia	Aves	Insecta
Order	Carnivora	Falconiformes	Hymenoptera
Family	Canidae	Falconidae	Apidae
Genus	*Canis*	*Falco*	*Apis*
Species	*Canis familiaris*	*Falco peregrinus*	*Apis mellifera*

Numerous adjectives have been used to qualify the word animal, some legally defined, others used more generally in this book and elsewhere. They include the following:

Domesticated — A term defined by the common law and Animals Act 1971 (see Chapter 2). It is also used by other legislation such as the Protection of Animals Acts. Generally, it includes species which have long lived in association with human beings for purposes such as food production or companionship. In biological parlance domesticated means captive-bred for many generations

Captive — An animal kept under some form of restraint (by man) varying from permanent caging to general supervision. The term is used in the Protection of Animals Acts (see Chapter 3)

Farm — A generalisation not used in legislation (*cf.* livestock) for animals commonly used in commercial agriculture, being an alternative to listing separate species when convenience is more important than precision

Enclosed — Usually used in respect of deer kept in a park; they are primarily fenced in but in some cases, especially in parks, may have some freedom to come and go

Farmed or ranched — Applied to deer, mink, foxes etc. kept for animal husbandry

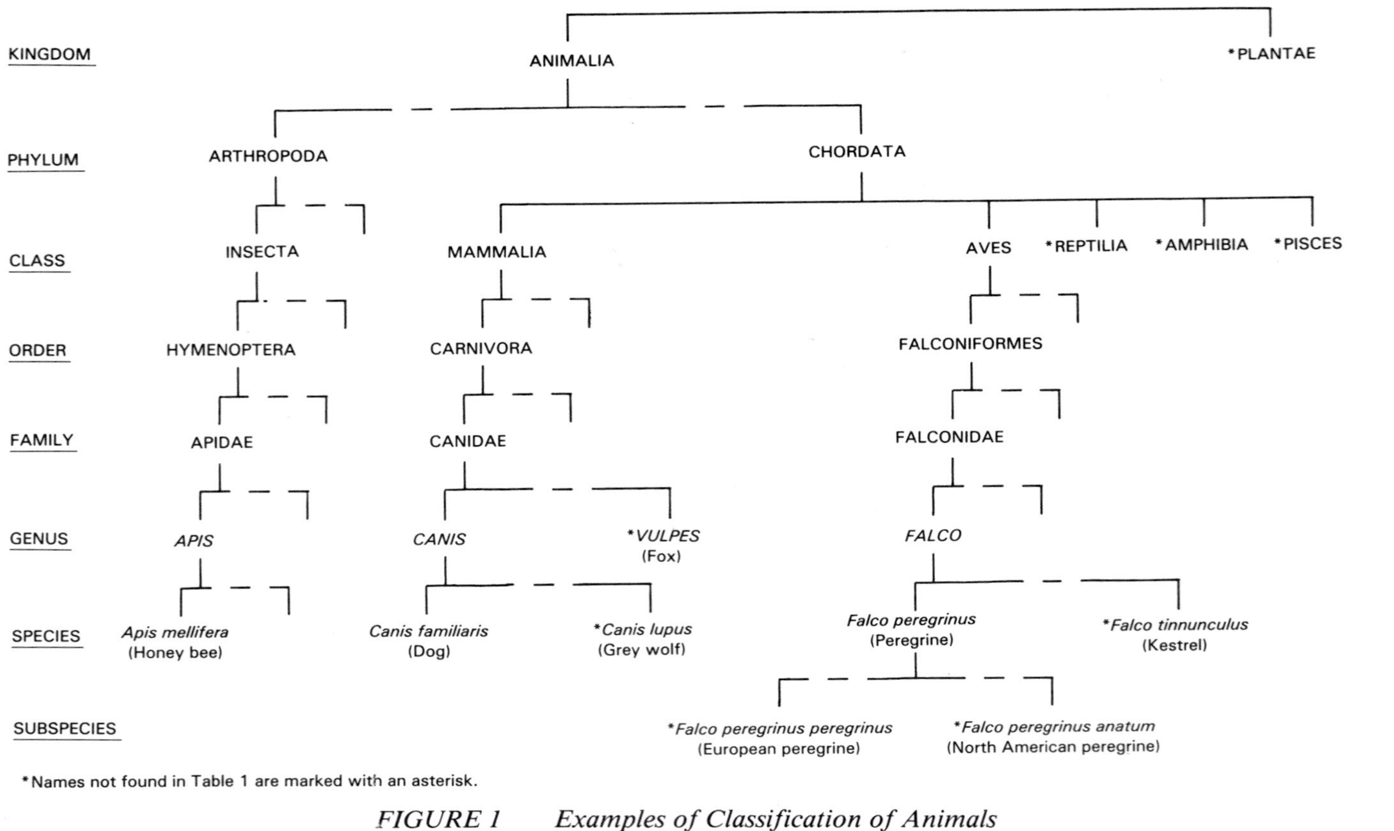

*Names not found in Table 1 are marked with an asterisk.

FIGURE 1 Examples of Classification of Animals

Protected	Applied in general to animals which have the protection of some form of law — usually conservation legislation. A specific term in the Animals (Scientific Procedures) Act 1986
Wild	Defined by the common law (see Chapter 2) and given specific meaning (as to wild birds and protected wild animals) in the Wildlife and Countryside Act 1981 (see Chapter 7). Although normally occurring as free-living animals, they may be found in captivity
Free-living	Referring to animals not living in captivity, this would normally relate to wild or exotic species (q.v.) but it may also be applied to domestic animals, although the adjective "feral" (q.v.) is often used. Thus a feral cat is a domestic cat (*Felis catus*) living independently of human control whereas the (Scottish) wild cat is a separate species (*Felis sylvestris*) and usually occurs as a free-living creature although it is also occasionally kept in zoological collections
Free-ranging	The same as free-living; preferred by those working with deer (DLC, 1984)
Feral	Animals (usually domesticated) living in a wild state after leaving captivity. Escaped wild animals are usually described as having reverted to the wild or a free-living state
Indigenous	A term, not used in legislation, referring to animals which have always existed in a given area
Exotic	Applied to non-indigenous animals, often excluding those already established in the wild in a particular country
Introduced	Non-indigenous animals which are established in the wild in areas in which they have not always existed

Endangered	Having some risk as to survival (also relates to species listed in the Endangered Species (Import and Export) Act 1976); may also be termed vulnerable or threatened. See Appendix 3 (Note 1)

ABBREVIATIONS

The abbreviations used in this book are given in the Table of Abbreviations and used hereafter without further explanation. The Ministry of Agriculture, Fisheries and Food is responsible for many aspects of animal health and welfare in England. Its counterpart in Wales is the Agriculture Department of the Welsh Office and in Scotland is the Department of Agriculture and Fisheries for Scotland. Purely for convenience and fluency the three bodies will be referred to as MAFF. The words Minister or Ministry should be construed likewise.

GENERAL READING

Certain works of reference are applicable to all or most chapters in this book and are of value for their breadth of information or particular relevance. To avoid their appearance in the list of references to general reading for each chapter they are listed here.

Readers should consult certain standard books of legal reference for further information and basic sources:

Halsbury's Laws of England, Vol. 2 and supplements (Halsbury, continuing, a)
Halsbury's Statutes, Vol. 2 and supplements (Halsbury, continuing, b)
Halsbury's Statutory Instruments, Vol. 2 and supplements (Halsbury, continuing, c)
Statutes in Force (HMSO, continuing, a)
Statutory Instruments (HMSO, continuing, b)

These books are usually available in the reference section of a good public library.

Literature on the law relating to animals which should be on the bookshelf of those interested in the subject or involved with animals are Cooper (1981), Crofts (1984), Kenny (1979), RCVS (1987), which is updated annually, and Sandys-Winsch (1984 a). To these may be added some of the more specialised books mentioned at the beginning of this chapter or referred to in subsequent chapters depending upon the reader's needs.

Excellent in their time, but now, sadly, out of date and out of print, are Field-Fisher (1964) and Thomas (1975). Bresler (1986) has published a collection of shallow, lighthearted accounts of cases involving animals. A useful introduction to general principles of English law is provided by Padfield (1985). For further insight into basic principles of law, reference may be made to professional or student textbooks although this should not be considered a substitute for obtaining advice from a qualified lawyer when the need arises.

FOREIGN LEGISLATION

Many areas of UK law are now affected by legislation made by the European Economic Community (EEC). Much UK legislation has been amended to implement EEC directives. In some fields EEC regulations have been produced; these take direct effect in UK law without the need for national enabling legislation. Since 1 January 1986 the members of the EEC are Belgium, Denmark, France, West Germany, Greece, Ireland, Italy, Luxembourg, The Netherlands, Portugal, Spain and the UK.

The Council of Europe has also produced a number of conventions which relate to animals, particularly in respect of welfare and conservation. Membership of the Council of Europe comprises the 12 EEC countries together with Austria, Cyprus, Iceland, Lichtenstein, Malta, Norway, Sweden, Switzerland and Turkey.

This European legislation and various international conventions affecting animals are not referred to in the general text since they have been implemented by national legislation, but they are discussed in Chapter 9.

Most countries outside the UK have their own legislation relating to animals, covering comparable subject matter (see Chapter 9).

TYPES OF LAW

The law can be categorised in various ways but certain aspects are referred to from time to time in this book.

Statute Law

This is the body of law contained in Acts of Parliament, which are also referred to as statutes. They are the prime source of legal authority for central and local government.

Subsidiary legislation, which is made by virtue of powers conferred by statute, includes statutory instruments, which are also called orders, regulations and Orders in Council. The last are made by the Privy Council (e.g. the Veterinary Surgeons (Practice by Students) Regulations Order in Council 1981 was so made under the authority of section 19(3) of the Veterinary Surgeons Act 1966).

Orders and regulations are normally made by a government minister. The Transit of Animals (General) Order 1973, for instance, was made by the Minister for Agriculture, Fisheries and Food and those for Wales and Scotland, by virtue of the Diseases of Animals Act 1950, which is now replaced by section 37 of the Animal Health Act 1981, and the Misuse of Drugs Regulations 1985 were made by the Home Secretary under section 10 of the Misuse of Drugs Act 1971.

Although statutory instruments are made by a minister they are subject to the approval of Parliament. This takes the form of either an "affirmative resolution" whereby the approval of both the House of Lords and the House of Commons is required for it to take effect or by "negative resolution" when an order comes into force after 40 days of its being "laid before Parliament" unless annulled by the resolution of one House. An order which is not properly authorised is invalid.

Local authorities (such as district or country councils) and other bodies (such as British Rail) are given power to make bye-laws by statute, such as those made under the Public Health Act 1936.

Common Law

This comprises the body of law built up by the courts or based on ancient custom.

There is some overlapping in this area since certain former common law principles have been replaced by similar statutory provisions, as in the Animals Act 1971 (see Chapter 2). Similarly, some law is a combination of common and statutory law; for example the sale of goods is based on both the common law principles of contract and the Sale of Goods Act 1979.

Criminal Law

The criminal law imposes obligations on people which are enforced by the state by way of punishment. The latter are usually a fine with or without imprisonment although, in respect of animal welfare offences, additional penalties may be imposed such as disqualification from holding certain licences to keep animals (see Chapter 3) or the confiscation of an animal or other equipment, as in the Wildlife and Countryside Act 1981 (see Chapter 7).

Prosecution is usually taken at the instigation of the police or some other official body. However, English (as opposed to Scottish) law also permits an individual person or organisation to bring a private prosecution; this is particularly used by animal welfare and conservation bodies to enforce legislation such as the Protection of Animals Acts 1911–1964 (see Chapter 3) or the Wildlife and Countryside Act 1981 (see Chapter 7). In the case of RSPCA *v.* Woodhouse (1984) it was held that this right of prosecution exists despite wording which appears to require the enforcement of orders made under the Animal Health Act 1981 to be carried out by local authorities.

Civil Law

The civil law deals with rights and duties as between individuals, including corporate bodies. They are enforced by a plaintiff bringing a claim in the County or High Court for damages (financial compensation) or for an injunction (a court order) requiring the defendant to cease or to remedy some fault.

The civil law, alternatively known as the law of tort, encompasses various heads of civil wrongs (including negligence, nuisance and trespass) and the law of contract, whereby commercial transactions are regulated.

While much civil law is based on case law, specific legislation is also applicable as in the Occupiers' Liability Acts 1957 and 1984 and the Sale of Goods Act 1979.

CASE

RSPCA *v.* Woodhouse (1984) C.L. 693.

REFERENCES

Blackmore, D.K. (1972). Some observations on the diseases of *Brunus edwardii* (species nova). *Veterinary Record* **90**, 382–385.

Bresler, F. (1986). *Beastly Law*. David & Charles, Newton Abbot.

Cassell, D. (1987). *The Horse and the Law*. David & Charles, Newton Abbot.

Cooper, J.E. and Cooper, M.E. (1982). *Naja rubika* — some veterinary problems in a new species. *Veterinary Practice* **20**, 4–5.

Cooper, M.E. (1981). *The Law for the Biologist*. Institute of Biology, London.

Crofts, W. (1984). *A Summary of the Statute Law relating to Animal Welfare in England and Wales*. Universities' Federation for Animal Welfare, Potters Bar.
DLC (1984). *Guidelines for the Safe and Humane Handling of Live Deer in Great Britain*. Deer Liaison Committee and Nature Conservancy Council, Peterborough.
Field-Fisher, T.G. (1964). *Animals and the Law*. Universities Federation for Animal Welfare, Potters Bar.
Gregory, M. (1974). *Angling and the Law*, 2nd edn. Charles Knight, London.
Halsbury (continuing, a). *Halsbury's Laws of England* (Lord Hailsham of Marylebone, ed.). Butterworth, London.
Halsbury (continuing, b). *Halsbury's Statutory Instruments*. Butterworth, London.
Halsbury (continuing, c). *Halsbury's Statutes*. Butterworth, London.
HMSO (continuing, a). *Statutes in Force*. HMSO, London.
HMSO (continuing, b).*Statutory Instruments*. HMSO, London.
Kenny, S. (ed.) (1979). *Animals*. Court 4, 4, 3–22.
North, P.M. (1972). *The Modern Law of Animals*. Butterworth, London.
Padfield, C.F. (1985). *Law made Simple*. W.H. Allen, London.
RCVS (1987). *Legislation Affecting the Veterinary Profession in the United Kingdom*, 5th edn with annual supplement. Royal College of Veterinary Surgeons, London.
Sandys-Winsch, G. (1984 a). *Animal Law*, 2nd edn. Shaw, London.
Sandys-Winsch, G. (1984 b). *Your Dog and the Law*. Shaw, London.
Thomas, J.L. (1975). *Diseases of Animals Law*. Police Publishing, London.
Weatherill, J. (1979). *Horses and the Law*. Pelham, London.
Williams, G.L. (1939). *Liability for Animals*. Cambridge University Press, Cambridge.

2 Responsibility for Animals

The tame animals, fuch as horfes, cows, fheep, and the like, are creatures, who by reafon of their fluggifhnefs and unaptnefs for motion, do not fly the dominion of mankind; but ufually keep within the fame paftures and limits, and may be eafily purfued and overtaken, if by accident they fhould efcape; and therefore the owner has the fame kind of property in them, as he has in all inanimate chattels.

("A Digest of the Law relating to the Game of this Kingdom", John Paul, 1775)

RESPONSIBILITY AND RIGHTS

The person who is responsible for an animal and who has rights in respect of it is usually its owner. Rights and responsibilities over an animal may also attach to those who do not have ownership in the fullest sense. Thus, a person may be responsible for an animal as its keeper, while it is in his possession or control, although the actual ownership is held by another person.

The law discussed in this chapter is primarily that applicable in England and Wales; Scottish law varies somewhat: that on civil liability for animals is set out in SLC (1982). See Appendix 3 (Note 2).

THE ROYAL PREROGATIVE

By virtue of the Royal Prerogative the Crown is entitled to various rights with respect to wild animals. This right is based on ancient custom but is described in early legislation and litigation which was initiated to protect that right. The Wild Creatures and Forest Laws Act 1971 abolished all such prerogative rights except those relating to royal fish and swans.

An Act of 1324 sets out the Crown's right as being "no general property . . . fish . . . except whales and sturgeon". Various authors have defined royal fish differently but, certainly, both in England and Wales as well as in Scotland the right only relates to the larger marine mammals of this kind (Fraser, 1977). See Appendix 3 (Note 3).

In 1592 the Case of Swans defined the right of the Crown thus: "as a swan is a royal fowl, and all those the property whereof is not known, do belong to the King by his prerogative". The case then distinguished between swans "swimming in open and common river" which are royal swans and those for the time being in privately-owned waters which belong to the owner thereof. Marked swans on the River Thames belong either to the Crown or the Dyers' or the Vintners' Companies. These birds are marked annually at the ceremony known as "swan upping".

Today the Department of Trade and Industry is responsible for the disposal of royal fish, and beached whales are reported to the Receiver of Wrecks, often via the Coastguard Service. The British Museum (Natural History) is consulted with regard to the acquisition of royal fish as museum specimens.

A stranded marine mammal is at risk from souvenir hunters and may require physical protection. In the case of Steele *v*. Rogers (1912) it was held that a beached whale was not a "captive animal" and that it was not, therefore, an offence of cruelty to cut pieces from it while still alive.

OWNERSHIP IN COMMON LAW

For the purposes of the common law an animal may be defined as any species other than a human being, although statutes very often have much more restricted definitions. Because of the wide variation in the meaning of the word "animal" in law (see Chapter 1), it is important to ascertain it anew in every situation.

The nature of ownership over animals in common law, which determines proprietary rights over an animal, depends upon the classification of the animals involved.

Animals are divided into domestic and wild (technically defined as animals *mansuetae* (or *domitae*) *naturae* and *ferae naturae* respectively). Domestic animals are those which have traditionally lived in association with mankind by virtue of their usefulness, e.g. for food or companionship. They include farm species and the cat and dog. It is for the courts to decide as a matter of fact whether an animal is domestic or not and a number of interesting cases have arisen. It has been held that in English law an elephant is not a domesticated animal (Filburn *v*. People's Palace & Aquarium Co. Ltd. (1890)). A camel, however, was held to be a domestic species on the grounds that camels are no longer found in the wild state (McQuaker *v*. Goddard (1940)) (but see Animals Act 1971, later).

Domestic Animals

Domestic animals can be owned with the same rights as in the ownership of inanimate goods. This principle is also reflected in statutes such as the Theft Act 1968 and the Sale of Goods Act 1979 (see later in this chapter).

Wild Animals

Wild animals form a residuary category of any animals not classified as domestic. They may include those not indigenous to Great Britain and indigenous species which are not domesticated, e.g. foxes, fish or wild birds.

It is a basic tenet of common law that wild animals are not property and, in their free-living state, cannot be owned. However, this rule has been modified in several ways to encompass situations in which man has exercised a limited measure of control over wild animals.

Reclaimed wild animals
Reclaimed wild animals are those which have been put under some restriction which may be total, as in the case of a caged bird or mammal, or partial, as with park deer, a trained falcon, racing pigeons or reared game birds. These animals are considered property while they are under restraint or, if not fully confined, while they remain with the owner or on his land or while they exhibit an intention to return.

Young wild animals
The owner of land owns the young of wild animals which are born on that land until they are capable of flying or running away (Blades *v*. Higgs (1865)).

Hunting rights
A landowner has the right to hunt, take or kill wild animals which are on his land at any given time. This right can be granted to other people and may pass automatically on granting a lease of the land unless specifically reserved. The carcass of a dead animal belongs to the owner of the land, or of the hunting rights over the land, on which the animal was killed. This rule is blurred, however, when an animal is pursued from one person's land to another.

OWNER'S RESPONSIBILITIES AND RIGHTS

An owner is liable to a greater or lesser extent for the damage caused by the animals which he owns, for nuisance (e.g. when they are too noisy) or for trespass (e.g. when they escape). He must also comply with statutory requirements relating to animals. Conversely, when he suffers loss or damage in respect of his animals he may look to the law for redress. For the different effects of criminal and civil law, see Chapter 1.

Criminal Law

Theft

The distinction between wild and domestic is also relevant in cases of theft. The Theft Act 1968 section 4(4) provides that a wild animal which has not been tamed or ordinarily kept in captivity, or the carcass of such an animal, cannot be stolen unless it has been or is being taken into the possession of another person.

Deliberate harm to an animal

Deliberate harm to an animal is actionable in civil law (see earlier); in addition, it is an offence under the Criminal Damage Act 1971 section 1 for a person, intentionally or recklessly without lawful excuse, to destroy or to damage an animal. This includes not only domesticated animals but also wild ones which have been tamed or which are ordinarily kept in captivity or under some form of control (s.10(1)(a)). Such damage may also involve a breach of the Protection of Animals Acts 1911–1964 (see Chapter 3).

Civil Law

Specific remedies can be enforced by or against the owner of an animal according to the circumstances. Action is brought in the civil law courts and the remedies are damages (compensation), an injunction (a court order) to prevent recurrence, or an order for the return of property.

Trespass

It is trespass to enter another person's land without his permission. The straying of animals on to land has given rise to much litigation and various principles have emerged:

Cats	An owner is not responsible for the trespass of a cat nor for any damage caused by it
Dogs	There is no liability for trespass unless the owner deliberately sends a dog on to another's property. The

owner is liable for damage to livestock caused by a dog when trespassing (Animals Act 1971)

Livestock	(i.e. cattle, horses, mules, hinnies, sheep, pigs, goats, poultry, including pigeons and peacocks, and enclosed deer). The owner is liable for damage caused by these animals when trespassing unless the trespass is due to the fault of the landowner or to a failure in a duty to fence the land or arises from animals straying in the course of their lawful presence on a highway (Animals Act 1971 ss. 4 and 5). See Appendix 3 (Note 4)
Wild animals	An owner is liable for damage caused by reclaimed wild animals but not for those over which he has no control, although if he attracts or amasses wild animals on his land he may be liable for damage which they cause to other land

Nuisance

A civil law nuisance may result from keeping animals in a manner which causes substantial disturbance to neighbours or a neighbourhood, e.g. through excessive noise or smell or by keeping too many animals together in an urban area.

Collecting or attracting excessive numbers of wild animals on land may lead to responsibility for a nuisance which they cause. Liability may also arise for nuisance caused by naturally occurring groups of wild animals if the landowner was, or should have been, aware of them, following the principle in Leakey *v*. National Trust [1980].

Negligence

An owner or person responsible for, or having control over, an animal is liable for harm (death, injury, damage to property) caused by that animal which he should reasonably have foreseen and which he failed to prevent. There is no liability for an unpredictable accident nor when the plaintiff consents to the risk of harm being caused; damages will be reduced to the extent to which the plaintiff contributed to the damage done.

For example, a person who negligently allows an animal to run into the road thereby causing a traffic accident is responsible for the injury or damage caused. If the person injured contributed to the accident on account of, say, careless driving or faulty brakes, he will not recover full compensation for his injury.

Wrongful interference with goods
Under the Torts (Interference with Goods) Act 1977 and the tort of trespass to goods a claim may be made for the return of an animal which has been wrongly taken. This may arise in respect of a stolen animal although this common law right or remedy is quite distinct from the Theft Act 1968. Under the 1977 Act compensation may be claimed for harm caused deliberately or negligently to the animal, e.g. if it has been injured, killed or infected. Practical examples would include damage caused by unauthorised veterinary treatment or even first aid.

Animals Act 1971
The Animals Act 1971, which has been examined in depth by North (1972), consolidated the previous common law rules regarding death, injury or damage caused by domestic animals which were often summarised in the saying "a dog is allowed its first bite". The Act provides that the keeper of a domestic animal is not liable for the damage or injury which it causes, provided that he was not aware of its propensity to cause such harm. Conversely, the keeper of a dangerous species is liable for any harm it does (s. 2).

The Act provides that the keeper of an animal is the owner or the person who has possession (which might be described as the day-to-day control) of it. This also includes the head of a household in which a person under 16 years old owns or possesses an animal. If possession is lost, e.g. in the case of an escaped animal, the person who was last its keeper is treated as remaining responsible for it whilst it is at liberty. A person who takes control of an animal to prevent it from causing damage or to return it to its owner is not treated as a keeper for the purposes of the Act (s. 5(3)).

The degree of liability for damage caused by an animal depends upon whether or not it is classified by the Act as a dangerous species. A dangerous species is defined (s. 6(2)) as:

(a) one which is not commonly domesticated in the British Islands; and
(b) whose fully grown animals normally have such characteristics that they are likely, unless restrained, to cause severe damage or that any damage they may cause is likely to be severe.

Any other species is considered not to be a dangerous species for the purposes of the Act.

The classification between dangerous and non-dangerous follows the pattern of the distinction between wild and domestic animals discussed earlier. The nature of individual animals is overridden by that of the species as a whole; thus elephants are considered to be non-domesticated (Behrens *v*. Bertram Mills Circus Ltd. [1957]) although, by comparison, in a case in Burma they were held to be domesticated (Williams, 1939). "Domesticated" means more than merely tamed (see Chapter 1) or that some of a species are kept as pets or used in agriculture. For example, the fact that some deer are farmed is probably not sufficiently widely and long enough established to fulfil the requirement ''commonly'' — although it must be admitted that they have been kept in a semi-wild state in parks for many centuries. Likewise the decision in McQuaker *v*. Goddard [1940] that a camel is a domesticated species is probably overruled by the Act.

Liability for dangerous animals is said to be ''strict''. In other words the keeper is liable for any harm caused by a dangerous animal even if he had no previous knowledge of its ability to do any particular damage.

Liability for all other animals which are not within the definition of dangerous animals (s. 2(2)) is restricted to liability for damage due to characteristics not normally found in such species and of which the keeper, or person looking after the animal for him, was already aware. The damage must be of a kind which the animal, unless restrained, would be likely to cause or which, if caused, was likely to be severe.

Liability is reduced to the extent either that it was caused by the fault of the person suffering the damage or that the person injured voluntarily accepted the risk of damage (e.g. by asking to enter an enclosure which he knew to contain dangerous animals (s. 5)). An employee working with such animals is expressly stated by the Act not to be accepting the risk of damage (s. 6(5)). Thus, he is not precluded from suing his employer for compensation if injured.

A trespasser cannot claim compensation for injury caused by an animal unless the animal was being used to protect persons or property in circumstances which could not be considered reasonable (s. 5(3)).

The Scottish law relating to harm done by animals was under revision in 1987. See Appendix 3 (Note 5).

Occupiers' Liability Acts 1957 and 1984

The occupier of property (including land, buildings and other structures) is responsible for taking reasonable care to ensure that those lawfully present there are not harmed. Thus, a person injured by a dog (Kavanagh *v*. Stokes [1942]) or, say, by animal cages which were structurally unsafe, may claim damages. However, if that person was a trespasser, the occupier is not liable (as in Murphy *v*. Zoological Society of London [1962] in which a person climbed into a lion's cage without permission). If the occupier is aware of some danger to trespassers he must under the Occupiers' Liability Act 1984 take reasonable steps to make it safe (Haley, 1984).

RESPONSIBILITY FOR DOGS

The Animals Act 1971 (s. 3) makes the keeper of a dog liable for livestock which it has killed or injured, namely cattle, horses, asses, mules, sheep, pigs, goats, poultry and captive deer, pheasants, partridges and grouse. This does not apply where, *inter alia*, the damage was due solely to the fault of the owner of the livestock.

A person who kills or injures a dog which was worrying livestock, and who had no other reasonable means to stop it, is not liable to the dog's owner provided that the former was authorised to act for the protection of livestock and reported the matter to a police station within 48 hours (Animals Act 1971 s. 9).

Worrying (attacking or chasing) livestock on agricultural land is an offence under the Dogs (Protection of Livestock) Act 1953 (see Chapter 3). This does not apply to a dog owner on whose land livestock are trespassing unless he deliberately causes the dog to attack them. A dog proved to have worried livestock may be treated as a dangerous dog under the Dogs Act 1871 section 2 (see later in this chapter).

Licences and Collars

By the Dog Licences Act 1959 the keeper of a dog must hold an annual licence for it unless it is a working sheep dog, a blind person's guide dog or a puppy under six months old. An increased licence fee was at one time under consideration, the income to be used to deal with the problems of dogs in the community — particularly strays (JACOPIS, 1985). However, it was announced in July 1986 that the licence and fee are likely to be abolished.

The Control of Dogs Order 1930 (as amended) requires that every dog must wear a collar whilst on a highway or in a public place. The collar must have the name and address of the owner on a plate or badge attached to it.

Stray Dogs

Under the Dogs Act 1906 the finder of a stray dog may return it to its owner or take it to the nearest police station. In the latter case either the finder chooses to keep the dog (for a minimum of one month) after obtaining a certificate of the details of the finding or the police keep the dog as if they had seized it.

The police may seize a stray dog and keep it for a minimum of seven days.

The animal must be returned to an owner who claims the dog and pays the costs of keeping it. If the police know the owner they must give him seven days' notice in which to claim the dog. After that time the police may humanely destroy the dog or they may sell it, other than for the purposes of animal research.

Dangerous Dogs

The Dogs Act 1871 (s. 2) permits a magistrates' court to order the owner of a dog to keep it under proper control or to have it destroyed. A complaint must have been made to the court that the dog is dangerous and not kept under proper control.

Bye-laws

Local authorities may produce bye-laws for the control of dogs — these may relate to keeping dogs on a leash, the prevention of fouling of pavements, the prevention of livestock worrying and the control of dogs in parks.

A local authority has power under the Road Traffic Act 1972 section 31 to designate roads on which dogs must be kept on a leash.

Guard Dogs

The Guard Dogs Act 1975 (s. 1) requires that a dog which is being used to protect premises must either be secured so that it cannot roam around the property or be kept under the control of a handler. At the entrance of the premises being guarded there must be a notice warning of the use of guard dogs.

ROAD ACCIDENTS

If cattle or a horse, ass, mule, sheep, pig, goat or dog is killed or injured in a road accident involving a motor vehicle other than that in which it is travelling the driver must stop. He must give his name and address, that of the owner of the motor vehicle involved and the registration number to any person having reasonable grounds for requiring the information. If this is not given, the accident must be reported within 24 hours to the police (Road Traffic Act 1972 s. 25). See also Chapter 3 and Appendix 3 (Note 6).

SALE AND PURCHASE OF ANIMALS

The sale and purchase of animals are subject to the same law as that for other goods, namely the principles of contract and the Sale of Goods Act 1979 and allied legislation. The whole topic is complex but the matters most likely to affect buyers and sellers are discussed here.

For a binding contract of sale there must be an offer which has been unconditionally accepted based on an intention to create legal relations between parties able in law to contract (e.g. they are not under 18 years of age). There must also be some "consideration" which is usually the price but is not necessarily the actual value of the animal or, indeed, a sum of money.

A contract for the sale of goods need not be in writing although the more valuable the animal being sold or the more complex the agreement the more important it is to have evidence of the agreement. Indeed, in situations where the same parties deal over many years terms may be built up in the course of correspondence or an understanding.

The parties to a sale of an animal are at liberty to include whatever terms they wish into their agreement. However, more often than not, except in the case of valuable pedigree animals, horses and farm animals, few matters will have been stipulated except for the price.

Such terms as are inserted into an agreement, verbally or in writing, are known as either conditions or warranties. If a term is of fundamental importance to the contract it becomes a condition; if it is peripheral then it will be a warranty. Much of the case law in this respect has been built up in respect of horses (Sophian, 1972; Weatherill, 1979; Cassell, 1987) and cattle. Generally, stipulations as to matters such as soundness and vice in horses or their equivalent in other species have been treated by the courts as warranties. However, it might be otherwise if the contract were more specific and involved, for instance, the supply of specially prepared laboratory animals for a particular piece of research.

The Sale of Goods Act 1979 supplements the general inadequacy of most day-to-day transactions by providing a structure dealing with matters such as the time for payment (i.e. on delivery) and the place of delivery (i.e. the seller's premises) or remedies when something goes wrong such as a failure to deliver or to pay for the animal.

The Act also implies certain terms in a sale of goods which are of particular benefit to a purchaser. A condition is implied that the seller has the right to sell the animal and a warranty that the buyer will enjoy "quiet possession" of it. Not to be taken literally, the latter means that the buyer can claim compensation if the animal in fact belongs to a third party who has recovered it from the purchaser.

If animals are sold by description rather than on sight (e.g. 12 gnotobiotic mice or a dog of specified sex, breed and pedigree) then those supplied must comply with the description.

Two very significant conditions are implied in specific circumstances which, if they are to be relied on in a dispute, make it important to identify the type of transaction and the circumstances in which it is made.

Whereas the foregoing terms (quiet possession and compliance with description) apply to all contracts the conditions commonly known as those for fitness for purpose and merchantable quality apply only when the seller is acting in the course of business (e.g. as a farmer, commercial breeder or pet trader) but not when an animal is being sold privately (e.g. the occasional sale of puppies from the family pet).

Furthermore, by virtue of the Unfair Contract Terms Act 1977, the trade seller may specifically exclude these conditions unless the buyer is a private customer (i.e. in what is known as a consumer sale) or he is trading on his written standard terms of business. In transactions with commercial concerns the seller may only exclude these conditions if it is reasonable to do so.

The condition for fitness for purpose implies that an animal will be reasonably fit for a purpose which the buyer expressly (e.g. when the animal is needed for a specific use in a laboratory or for breeding stock) or impliedly (where it is clear from the surrounding agreement that the animal is wanted, for example, for a family pet) makes known the purpose for which it is required. The buyer must also have relied on the skill and judgment of the seller.

The condition for merchantable quality provides that the animal sold will be as fit for the purpose for which such animals are commonly purchased as it is reasonable to expect in the circumstances, including the description applied to them and the price. Thus, to use an analogy from ordinary shopping, cheap goods described as secondhand or "seconds" quality are not expected to be as sound as new and perfect items sold at the full price.

In respect of this condition a buyer cannot complain of any defects which were pointed out to him before the sale or which, if the buyer examined the animal, he should have found on examination.

The moral of these two conditions is that a buyer either should buy entirely "blind" and rely on the conditions if the animal turns out not to be of the correct quality or should have the animal examined by a skilled person prior to purchase.

The conditions for fitness for purpose and merchantable quality can give rise to difficulties of interpretation and of their not being implied into every transaction; besides which, some authorities argue that there is no condition of merchantable quality implied in a sale of animals. Consequently, in all but the most insignificant sales the purchaser should insist on written terms of agreement, including conditions or warranties as to the quality of the animals supplied. See also Appendix 3 (Note 7).

The average purchaser is not concerned about the terms of his agreement unless a problem arises, e.g. failure or delay of delivery or, most commonly, animals which do not meet his expectations. In the circumstances what are the remedies? In the first place, if there is a breach of a condition there is the right to return the animal and recover the price plus compensation for any consequential expense or loss involved (telephone, postage, travel or bank charges or loss of other contracts). However, this right is very easily lost, particularly if the animal is not rejected promptly or if the buyer does anything to suggest that he has accepted it, e.g. by not informing the seller of his complaint or by selling the animal to another person.

In such circumstances the Sale of Goods Act provides that the purchaser must treat the implied terms as warranties, not conditions, i.e. he must keep the animal and accept compensation for failure to fulfil the term.

A purchaser has the option of treating a condition as a warranty. Therefore it is open to a purchaser of an animal which turns out to be substandard (e.g. subject to some disease, injury or congenital defect which was not apparent at the time of purchase) to accept veterinary treatment or a replacement in lieu of returning the animal or compensation. However, acceptance of veterinary treatment should be without prejudice to the purchaser's basic rights in case therapy does not correct the problem.

A buyer has other rights which are not dependent on express or implied conditions. If he was misled by misrepresentation by the seller he has a right to return the animal and/or to compensation under the Misrepresentation Act 1967. An untrue statement of fact must be made by the seller in the course of negotiations which induced the buyer into the agreement. The buyer must complain very promptly if he wishes to return the animal, for if he delays or does anything inconsistent with the seller's ownership, such as reselling the animal or breeding from it, the remedy is simply financial compensation. Provided no fraud is involved the seller can defend his statement by proving that he genuinely believed that it was true at the time of making the contract. If fraud is involved (deliberately incorrect descriptions of animals, for example, as to the sex, breed, fertility or soundness) this can lead to prosecution under the Trade Descriptions Act 1968 and 1972. The buyer has an economical way of recovering his loss in such situations since, in the event of a court action, he may ask for an award of compensation, thereby avoiding a private claim in the County Court.

The seller of an animal has remedies against the buyer who does not fulfil his part of the contract. If the latter has accepted the animal but not paid, the seller may sue him for the price.

If the buyer has not accepted the animal, the seller may claim damages; if he can resell the animal easily his compensation will be nominal but if there is no ready market or if he cannot make the same profit he is entitled to the value of his loss.

The unpaid seller has a right to resell an animal provided that he expressly reserved that right in his contract or gives notice to the buyer.

MISCELLANEOUS MATTERS OF OWNERSHIP

Gifts to Animals

An animal has no standing in a court of law since it is not a person. Likewise, it is not possible for an animal to own property or to receive a gift in its own right. However, one can make a gift to an individual on condition that it is applied for the benefit of the animal. This may be done by way of a trust made in the donor's lifetime or by will, to take effect on the donor's death.

Gifts may be given to animal charities, or one may be set up on condition that an individual animal be looked after. It is not possible to create a charity for the benefit of an individual animal since an element of benefit to the community in general is required to establish a charitable intention. Nevertheless, many charities exist to help animals in general (e.g. the Royal Society for the Prevention of Cruelty to Animals) or those of particular species (e.g. a rest home for horses).

Insurance

The importance of insurance should not be overlooked. The conscientious owner or keeper should consider cover for liability in respect of injury or damage caused by, or in the use of, his animal. If the animal is valuable it should be insured against theft or other diminution of its value. Insurance to cover veterinary fees is widely available. Those who keep animals in circumstances likely to attract opposition to their exploitation (notably in hunting, zoos or research) are now considering cover for the legal costs involved in defending prosecutions under the welfare legislation as well as additional cover for damage to people or property.

CASES

Behrens *v*. Bertram Mills Circus Ltd. [1957] 1 All E.R. 583.

Blades *v*. Higgs (1865) 11 H.L. Cas. 621.
Case of Swans (R. *v*. Lady Joan Young) (1592) 7 Co. Rep. 15b.
Filburn *v*. People's Palace & Aquarium Co. Ltd. (1890) 25 Q.B.D. 258.
Kavanagh *v*. Stokes [1942] I.R. 596.
Leakey *v*. National Trust [1980] 1 All E.R. 17.
McQuaker *v*. Goddard [1940] 1 All E.R. 471.
Murphy *v*. Zoological Society of London [1962] C.L.Y. 68.
Steele *v*. Rogers (1912) 76 J.P. 150.

REFERENCES

Cassell, D. (1987). *The Horse and the Law*. David & Charles, Newton Abbot.
Fraser, F.C. (1977). Royal fishes: the importance of the dolphin. In *Functional Anatomy of Marine Mammals* (Harrison, R.J., ed.). Academic Press, London.
Haley, M.A. (1984). The uninvited entrant and the Occupiers' Liability Act 1984. *Law Society's Gazette* **81**, 1594–1595.
JACOPIS (1985). *Dog Licensing: Future Developments in Great Britain*. Joint Advisory Committee on Pets in Society, London.
North, P.M. (1972). *The Modern Law of Animals*. Butterworth, London.
SLC (1982). *Consultative Memorandum No 55 on Civil Liability in Relation to Animals*. Scottish Law Commission, Edinburgh.
Sophian, T.J. (1972). *Horses and the Law*, 2nd edn. J.A. Allen, London.
Weatherill, J. (1979). *Horses and the Law*. Pelham, London.
Williams, G.L. (1939). *Liability for Animals*. Cambridge University Press, Cambridge.

RECOMMENDED READING

See Chapter 1 for literature generally applicable.

Macrory, R. (1982). *Nuisance*. Oyez Longman, London.
North, P.M. (1972). *The Modern Law of Animals*. Butterworth, London.
Rogers, W.H.V. (1984). *Winfield and Jolowicz on Tort*, 12th edn. Sweet & Maxwell, London.
Sandys-Winsch, G. (1984). *Your Dog and the Law*. Shaw, London.
Williams, G.L. (1939). *Liability for Animals*. Cambridge University Press, Cambridge.
Wollen, N.J. (1982). A lawyer's view of breeding agreements in zoos. In *Paignton Zoo Annual Report 1981*, Appendix 8. Paignton Zoo, Paignton.

3 *Welfare Legislation*

OFF THE BRUITE CREATURE

92. No man shall exercise any Tirranny or Crueltie towards any bruite Creature which are usuallie kept for man's use.
93. If any man shall have occasion to leade or drive Cattel from place to place that is far of, so that they be weary, or hungry, or fall sick, or lambe, It shall be lawful to rest or refresh them, for a competent time, in any open place that is not Corne, meadow, or inclosed for some peculiar use.

("The Body of Liberties", Massachusetts Bay Colony, 1641)

Samedi, le 25e Mai, 1811 . . . Sur la représentation faite à la Cour par les Officiers du Roi que des Enfants presque journellement font voler des Hannetons (cockchafers or maybugs), en leur attachant un fil avec une Epingle, — LA COUR, afin de mettre fin à une coutume aussi cruelle, ouïe la conclusion des Officiers du Roi, a défendu d'en faire voler de la manière susdite, sur la peine de Trois Livres Tournois d'amende . . .

("Recueil d'Ordonnances de la Cour Royale de l'Isle de Guernesey", Tome Deuxième, 1801–1840)

The welfare of animals is the concern of numerous Acts of Parliament. The main body of them, which relate to cruelty, are the Protection of Animals Acts 1911–1964 (reference to which in the text includes, as applicable, the Protection of Animals (Scotland) Act 1912). Other statutes are directed at the care of animals in specific situations. The Animals (Scientific Procedures) Act 1986, which regulates the use of animals in research, is dealt with separately in Chapter 4.

LEGISLATION AGAINST CRUELTY

Protection of Animals Acts 1911–1964

Protected animals

The Protection of Animals Acts apply to any domestic or captive animal. The former category includes animals commonly considered domesticated such as horses, cattle, cats, dogs and birds as well as any other species which has been "sufficiently tamed to serve some purpose for the use of man" (s. 15). The term "captive" covers any other species, including any bird, fish or reptile, which is in captivity or under some form of control (such as caging or pinioning) to keep it confined. The captivity must be something more than temporary prevention from escape or inability (not caused by man) to escape. Thus in two cases, Steele *v*. Rogers (1912) (a whale stranded on the beach) and Rowley *v*. Murphy [1964] (a hunted deer restrained whilst it was killed), it was decided that the animal involved was not captive within the meaning of the Act. These cases were followed by the court in Hudnott *v*. Campbell (1986), a decision involving deliberate injury to a hedgehog (Muriel, 1986a).

It follows that free-living non-domesticated animals are not covered by the Act until they are taken into captivity. However, many such species have some protection under conservation legislation, in particular the Wildlife and Countryside Act 1981 (see Chapter 7). See Appendix 3 (Note 8).

Cruelty

Section 1 of the main Acts makes it illegal cruelly to ill treat or, being the owner, to permit ill treatment of an animal in specified ways, e.g. by beating, terrifying or overloading it.

In addition, section 1 provides a more general offence of wantonly or unreasonably causing unnecessary suffering to an animal. This is an open-ended category which includes not only overt acts but also omissions such as the failure to provide necessary food, water or veterinary attention.

It was held in the case of Barnard *v*. Evans [1925] that the adverb "cruelly" is to be equated with "so as to cause unnecessary suffering". In order to show that an offence has occurred under section 1 it is necessary to prove both that the act caused suffering and that the suffering was unnecessary (Ford *v*. Wiley (1889)) and substantial (Swan *v*. Saunders (1881)). The words "wantonly or unreasonably" were described in the New Zealand case McEwan *v*. Roddick [1952] as being "the callous disregard for or a reckless indifference to the suffering of an animal". For a discussion of the legal aspects of cruelty see Hill (1984).

Various other matters are specifically made offences of cruelty punishable under section 1 of the Act, notably:

(a) The transportation of an animal in a way which causes it unnecessary suffering
(b) Animal fighting and baiting
(c) The deliberate poisoning of an animal without reasonable cause
(d) Operations which are performed on an animal without due care and humanity

(e) The failure of the owner of an animal to exercise reasonable care and supervision over it so as to protect it from cruelty

Poisoning

In addition to the illegality of administering poison to an animal it is an offence to sell or supply poisoned grain other than for use in agriculture (s. 8(a)) or to put poison on land or in buildings unless it is used to destroy small vermin and all reasonable precaution is taken to prevent access to it by domestic animals (s. 8(b)). The Animals (Cruel Poisons) Act 1962 forbids the use of poisons known as red squill and yellow phosphorus altogether and the use of strychnine to kill mammals of any kind except moles.

The Wildlife and Countryside Act 1981 sections 5(1) and 11(2) contains provisions preventing the use of poisonous, poisoned or stupefying substances to kill Schedule 6 wild animals and the Game Acts 1831 and 1970 forbid the laying of poison with intent to destroy or injure game (see Chapter 7, Table 1).

The use of poison or poisonous gas is permitted in certain circumstances for the control of pests (see Chapter 7).

Road accidents

The 1911 Act provides for situations (such as road accidents or grave neglect involving horses, mules, bulls, sheep, goats and pigs) when the owner is not available or is uncooperative. The police have power to authorise, on the basis of a veterinary surgeon's certificate, the movement or destruction of an animal which is so sick or injured that it would be cruel to keep it alive.

Penalties

The Act (and subsequent legislation) provides for enforcement in the form of fines and imprisonment but additional powers are available to a court:

(a) To order the humane destruction of an animal if it would be cruel to keep it alive (s. 2)
(b) To order an animal to be taken away from a person convicted under the Act if it is likely to be exposed to further cruelty; the court can dispose of the animal as it thinks fit (s. 3)
(c) Following a conviction, to order disqualification from keeping a dog and holding a dog licence (Protection of Animals (Cruelty to Dogs) Act 1933, Protection of Animals (Cruelty to Dogs) (Scotland) Act 1934)
(d) Following a person's second or subsequent conviction for cruelty, to disqualify him from keeping all or any specified animals for such period as it thinks fit (Protection of Animals (Amendment) Act 1954)

(e) To order up to £1000 compensation for damage or injury done to person, property or animal (s. 4)
(f) Following a conviction under the Protection of Animals Acts, to order disqualification from holding or obtaining a licence issued under the Pet Animals Act 1951, the Dangerous Wild Animals Act 1976 and other similar legislation. See Appendix 3 (Note 9).

Abandonment of Animals Act 1960

The Abandonment of Animals Act 1960 provides that it is an offence of cruelty under the 1911 Act to abandon an animal (as defined in the main Act) without reasonable excuse in circumstances likely to cause it unnecessary suffering. Although the Act was designed to apply to deserted pets it is also applicable to the release of other species, such as rehabilitated wild animals or those bred in captivity, for the purpose of re-stocking wild populations. Thus a careful estimation, including physical condition, the environment and food supplies, should be made of an animal's ability to survive in the wild before its release (Cooper *et al.*, 1980).

Protection of Animals (Anaesthetics) Acts 1954 and 1964

The Protection of Animals (Anaesthetics) Acts 1954 and 1964 make it illegal to carry out on an animal an operation involving its sensitive tissue or bone structure without the use of anaesthesia sufficient to prevent its feeling pain. Failure to do so is deemed to be an operation performed without due care and humanity which is a specified offence under the 1911 Act. There are a number of exceptions to this provision, in particular:

(a) Giving first aid in an emergency to save life or relieve pain
(b) Hollow needle injections or extractions
(c) Minor procedures which either would not normally be carried out by a veterinary surgeon or would normally be carried out by a veterinary surgeon without anaesthesia
(d) Certain procedures on agricultural animals and the docking of the tail and amputation of the dew claws of a dog before its eyes are open
(e) Any experiment duly authorised under the Animals (Scientific Procedures) Act 1986 (see Chapter 4)

Deer antlers in velvet must be removed under anaesthesia sufficient to prevent pain unless the circumstances fall into exceptions (a) or (e) above (Removal of Antlers in Velvet (Anaesthetics) Order 1980).

The Protection of Animals (Anaesthetics) Acts 1954 and 1964 expressly do *not* apply to "a fowl or other bird, fish or reptile". Thus the failure to use anaesthesia for an operation on these species is not automatically deemed to have been carried out without "due care and humanity" which is a specific offence of cruelty under the 1911 and 1912 Acts. Nevertheless, a successful prosecution might be achieved by proving that in the particular circumstances of the case failure to use anaesthesia in a bird, fish or reptile amounted to a lack of due care and humanity or, alternatively, that it constituted unnecessary suffering.

Cinematograph Films (Animals) Act 1937

It is an offence to exhibit to the public, or to supply for exhibit, a film produced in Great Britain or elsewhere if any scene in the film was organised or directed so as to involve the cruel infliction of pain or terror or the goading of an animal to fury (Cooper and Cooper, 1981). See also WSPA (1981) and Wilkins (1987).

Protection of Animals Act 1934

The Protection of Animals Act 1934 makes bullfighting and many aspects of rodeos illegal within Great Britain.

OTHER LEGISLATION RELATING TO WELFARE

Wildlife and Countryside Act 1981

The Wildlife and Countryside Act 1981 provides legal protection for non-domesticated species for the most part free-living and indigenous to Great Britain; this is considered in detail in Chapter 7.

Section 8, however, deals with captive birds and requires that a cage or other receptacle in which a bird is kept must be of sufficient dimensions to enable the bird to stretch its wings freely. The case of Starling *v.* Brooks [1956] held that confinement is a question of fact and a valid motive or absence of suffering is not a defence to a prosecution under section 8.

The section does not apply to:

(a) Poultry
(b) A bird which is being transported
(c) A bird which is being shown in a public exhibition or competition and

is confined for not more than 72 hours in aggregate (an open general licence has been issued permitting limited confinement in a show cage in order to accustom a bird to it for the purposes of exhibition or competition)

(d) A bird which is undergoing examination or treatment by a veterinary surgeon. The last exception does not apply, however, to care given by lay people in which case the cage, even if a hospital cage, must comply with section 8

Docking and Nicking of Horses Act 1949

It is illegal to dock or nick the tails of horses and other equids unless this is carried out for reasons of health, disease or injury and is certified as necessary by a member of the Royal College of Veterinary Surgeons.

Dogs (Protection of Livestock) Act 1953

The owner or person in control of a dog commits an offence if the dog worries livestock (cattle, sheep, goats, swine, horses and poultry) on agricultural land (arable, meadow or grazing land and that used for poultry etc.). The worrying of livestock includes attacking or chasing in a way which is likely to cause abortion, injury, suffering or loss or a diminution in their produce, or even (subject to exceptions for working dogs) simply being in a field with sheep unless the dog is on a leash or under close control (see Chapter 2).

Dogs Act 1906

Section 1(4) provides that a dog found to be worrying livestock may be treated by a court as a dangerous dog and made subject to a control or destruction order (see Chapter 2).

FARM ANIMAL WELFARE

Welfare on the Farm

Protection is afforded to farm animals by the Agriculture (Miscellaneous Provisions) Act 1968 section 1 whereby it is an offence for a person to cause unnecessary pain or unnecessary distress to livestock kept under his control on agricultural land. It is also illegal for such a person knowingly to permit (rather than to cause) such suffering.

The definition (in section 8) of livestock includes any creature kept for the production of food, wool, skin or fur or for use in the farming of land or for such purpose as MAFF may specify. The Welfare of Livestock (Deer) Order 1980 provides that deer kept for the production of antlers in velvet are to be included in the definition of livestock. Agricultural land is defined as that used in commercial agriculture.

The Act (s. 2) provides for the publication of regulations and codes of practice for the welfare of livestock kept on agricultural land. The following Regulations have been produced:

> The Welfare of Livestock (Intensive Units) Regulations 1978 make provision for the maintenance of automatic equipment and the prevention of unnecessary pain or distress in livestock in intensive care units
>
> The Welfare of Livestock (Prohibited Operations) Regulations 1982 prohibit the performance of a number of operations on livestock which are kept on agricultural land unless they are carried out as emergency first aid to save life or to relieve pain, as proper treatment by a veterinary surgeon for disease or injury or as a lawful act under the Animals (Scientific Procedures) Act 1986. See Appendix 3 (Note 10).

The prohibited operations refer largely to procedures used in animal production such as short tail docking in sheep and hot branding of cattle but some may also be employed in circumstances outside commercial agriculture, e.g. the de-voicing of cockerels or surgical castration of birds, operations more severe than feather clipping to impede the flight of birds and the removal of deer antlers which are in velvet. Since these Regulations relate only to livestock on agricultural land the procedures listed therein may be performed on animals in other situations provided that they do not infringe other legislation, particularly the Protection of Animals Acts. Thus, surgical pinioning may be performed on ornamental waterfowl; however, the removal of deer antlers in velvet must be carried out under anaesthesia and by a veterinary surgeon in any circumstances other than by way of first aid in an emergency or as part of an experiment authorised under the Animals (Scientific Procedures) Act 1986. The whole range of mutilations in animals has been examined by a working party of the RCVS with a view to assessing their severity and control.

Codes of recommendations (MAFF, various) or practice (MAFF, undated) have been published by MAFF in respect of cattle, pigs, sheep and poultry and for the air transportation of horses and other farm species. These codes have a limited legal status in that a failure to comply with a code cannot give rise to legal proceedings but may be used in a prosecution under section 1 of the 1968 Act as evidence of the defendant's breach of that section.

The Farm Animal Welfare Council is an independent body which advises the Minister of Agriculture on the welfare of animals. It has produced reports on a variety of welfare issues including the welfare of farmed deer (FAWC, 1985a) and the slaughter of livestock including religious methods (FAWC, 1984, 1985b), and has also provided advice to MAFF on codes for rabbits and ducks and revised draft versions of several existing codes. It has reviewed egg production systems (FAWC, 1986a) and made proposals for regulations which would strengthen the present codes (FAWC, 1986b). It is studying goat and sheep welfare and the transportation of livestock.

Slaughter

Killing an animal is not an offence under the 1911 Act provided that it is performed humanely, although it might be so if it were accompanied by unnecessary suffering or an act of cruelty. Killing an animal may constitute an offence under other legislation such as the Criminal Damage Act 1971 (see Chapter 1), the Wildlife and Countryside Act 1981 (see Chapter 7) or the Slaughterhouses Act 1974 (see later). The event may also involve a violation of the owner's rights in civil law (see Chapter 2).

Slaughter of Animals Act 1974

The Slaughter of Animals Act 1974 requires that horses, cattle, sheep, swine and goats which are to be killed at a slaughterhouse be killed instantly by a mechanically operated instrument in good repair, be instantaneously stunned by certain methods so that the animal is insensible to pain until death supervenes or be killed by other permitted methods such as those provided by the Slaughter of Pigs (Anaesthesia) Regulations 1958. These requirements do not apply to Muslim or Jewish ritual slaughter provided that it is performed without unnecessary suffering and in an approved manner; nor are they applicable to slaughter arising from an accident or emergency necessitating the prevention of physical injury or suffering to any person or animal.

Slaughter of Poultry Act 1967

As amended by the Animal Health and Welfare Act 1984, the effect of the Slaughter of Poultry Act 1967 is that slaughter, for whatever purpose, of domestic fowls, geese, ducks and guinea-fowls must be carried out by one of the methods specified in the Act, e.g. decapitation or dislocation of the neck or by stunning prior to death. There are exceptions for Jewish and Muslim ritual slaughter and the provision does not apply to animal research (see Chapter 4) or to a veterinary surgeon in the exercise of his profession.

These Acts and subsidiary legislation (Slaughter of Poultry (Humane Conditions) Regulations 1984 and the Slaughter of Animals (Prevention of Cruelty) Regulations 1958, amended 1984) deal with slaughterhouses and the environment in which animals are kept pending slaughter.

Markets

The welfare of cattle, sheep, goats and swine in markets is provided for in the Markets (Protection of Animals) Order 1964 (as amended) and the Poultry (Exposure for Sale) Order 1937. The orders require the provision of food and water, the penning of animals without injury and the inspection by a veterinary inspector with power to remove or to treat an animal which has been or is likely to be caused unnecessary suffering.

WELFARE IN TRANSPORT

The Animal Health Act 1981 (ss. 9, 37 and 38) provides for orders to be made regulating the welfare of animals while they are being transported either within, or to and from, the UK. The relevant orders, species and methods of transport to which they relate are set out in Table 1.

These orders have many provisions in common which are summarised below. However, while the 1973 Order speaks in general terms the others deal with such matters in greater detail:

(a) Animals must be fit to travel
(b) Vehicles and containers must be suitably adapted and free from potential causes of injury
(c) Adequate and suitable water, food, temperature, ventilation and humidity must be provided
(d) Personal attendance must be available when necessary
(e) The animals must be rested before, during and after the journey as necessary
(f) Responsibility lies with the carrier or any person in control of the animal during the journey or during loading or unloading

The legislation has been tested in the courts with varying degrees of success. Muriel (1985, 1986b), and Muriel *et al.* (1985) have drawn attention to some of its shortcomings. A review of the transport orders has been promised, commencing with the Poultry Orders (Hansard, 1986).

TABLE 1

Order	*Species*	*Method of transport*
Conveyance of Live Poultry Order 1919	Poultry*	All except air
Transport of Animals Order 1927	Farm**	Coastal waters
Animals (Sea Transport) Order 1930	Farm**	High seas
Transport of Animals (Road and Rail) Order 1975	Farm***	Road, rail
Transit of Animals (General) Order 1973	Mammals (except man) Other four-footed animals Fish, reptiles, crustaceans Other cold-blooded animals EXCEPT: species covered by Orders above	Any including air EXCEPT: Any covered by Orders above

*Domestic fowls, turkeys, geese, ducks, guinea-fowls and pigeons.
**Cattle, sheep, goats and all other ruminating animals and swine.
***Cattle, sheep, goats, swine and horses.

Article (7)1 of the Transit of Animals (General) Order makes it an offence to permit an animal which is ''unfit'' or likely to give birth on the journey to be carried. If this is to be construed literally, sick or injured animals cannot be transported, even to the veterinary surgery, for treatment. It has prevented a public carrier from accepting sick hedgehogs for conveyance to a wildlife hospital. See Appendix 3 (Note 11).

The Transit of Animals (Road and Rail) Order 1975 forbids the movement of an unfit (i.e. infirm, diseased, ill, injured or fatigued) farm animal or horse if it is likely to undergo unnecessary suffering. This provision is particularly relevant to the movement of animals which become sick or are injured on the farm. In such circumstances transportation of the animal to a slaughterhouse rather than slaughter at the farm may also constitute a breach of the Agriculture (Miscellaneous Provisions) Act 1968 and an offence of carrying an animal so as to cause it unnecessary suffering under the Protection of Animals Act 1911 (Anon, 1986).

It has been established in British Airways Board *v*. Wiggins [1977] that, in respect of animals covered by the 1973 Order, the carrier of, and any other person having charge of, an animal is responsible throughout its journey for the provision of a suitable receptacle for it.

In Air India *v*. Wiggins [1980] the House of Lords held that a carrier of animals which is in breach of the 1973 Order (e.g. an airline carrying animals in unsuitable containers or in conditions which cause unnecessary suffering) when it enters British territory may be prosecuted even if the carrier is a foreign national and its journey commenced overseas. However, no prosecution can be sustained in respect of animals which have died before entry into the jurisdiction.

The International Air Transport Association (IATA) has issued standards (IATA, annual) for the transportation of animals which are mandatory for member airlines. MAFF has produced a code of practice for the air carriage of farm animals (MAFF, undated), and recommended conditions applicable particularly to endangered species are to be found in CITES (1980).

Other carriers, such as British Rail, have regulations regarding the carriage of animals and the Post Office permit the posting of certain live insects only with prior approval (Post Office, annual). See Collins (1987).

WELFARE OF ANIMALS TO BE EXPORTED

The Animal Health Act 1981 (ss. 40–42) restricts the export of horses and ponies and makes provisions for the welfare of those which are exported. The Export of Horses (Protection) Order 1969 and allied Orders restrict the permitted size of horses leaving the country and provide for rest periods, veterinary inspection and certification of fitness prior to exportation. There are exemptions for thoroughbred horses exported for sporting, exhibition or breeding purposes.

The Act (s. 12) also provides powers to control the export of cattle, goats, sheep or swine for welfare purposes and the Export of Animals (Protection) Order 1981 requires rest periods and the issue of a veterinary certificate of fitness prior to export and lays down general conditions for accommodation and other matters during transportation.

LICENSING IN RESPECT OF ANIMALS KEPT FOR COMMERCIAL PURPOSES

Legislation exists to regulate most situations in which animals are kept in captivity for commercial purposes other than agriculture. This consists primarily of the licensing of persons by whom the animals are kept. It is carried out by the local authority in whose area the animals are maintained. The legislation also provides for the inspection of such premises by an officer of the local authority who may be, or in some cases must be, a veterinary surgeon. Provision is also made that the premises should comply with general standards, which are common to all but guard dogs and zoos. They require that the accommodation and environment must be suitable, that appropriate food, water and attendance are provided and that precautions are taken against the spread of disease, fire and other emergencies. A conviction for an offence against the cruelty or licensing legislation leads to refusal or loss of licence. Guides for inspectors under the relevant statutes have been produced by the British Veterinary Association.

The Acts and the situations to which they apply are as follows:

Breeding of Dogs Act 1973: premises where more than two bitches are kept with the intention of breeding for sale

Animal Boarding Establishments Act 1963: premises which are used for the boarding of other people's dogs and cats on a commercial basis, unless the accommodation is provided as ancillary to another business such as a veterinary practice or dog breeding concern

Riding Establishments Acts 1964 and 1970: any establishment where horses are kept for the purpose of hiring out for riding or for use for riding instruction for payment. Inspection must be carried out annually be a veterinary surgeon chosen from a list drawn up by the Royal College of Veterinary Surgeons (Sophian, 1972; Weatherill, 1974; Cassell, 1987)

Guard Dogs Act 1975: kennels provided commercially for guard dogs. That part of the Act requiring such kennels to be licensed is not yet in force. The remaining part requires that guard dogs, while protecting commercial premises, must be either kept under control by a handler or prevented from roaming freely on the premises

Pet Animals Act 1951: any business of selling animals (vertebrates) as pets (i.e. animals to be used for domestic or ornamental purposes). This includes not only the conventional high street pet shop but also trade from other premises. For example, the case of Chalmers *v*. Diwell [1976] held that the holding of birds at a private home prior to export for trade should have been licensed under the Pet Animals Act. The Act does not apply to a person running a business of breeding and selling pedigree animals nor to the sale of offspring of pets by their owner

Performing Animals (Regulation) Act 1925 and Performing Animals Rules 1925 and 1968: a person who exhibits at a public entertainment, or trains, any performing animal (vertebrate) must be registered with a local authority

Dangerous Wild Animals Act 1976 and Dangerous Wild Animals Act 1976 (Modification) Order 1984: the keeping, i.e. possession (which may be distinct from ownership), of an animal listed in the Schedule to the Act. A new

Schedule has been inserted by the 1984 Order and it includes many non-domesticated species which might be considered capable of causing greater injury than domestic animals.

No licence is required for scheduled animals whilst they are kept in a circus or premises registered as a zoo, a pet shop or under the Animals (Scientific Procedures) Act 1986. There is no specific exemption, beyond the last mentioned, for scientific or educational establishments *per se*; once an animal leaves exempted premises its keeper must obtain a licence. A licence is not required for a person who keeps an animal while it is undergoing veterinary treatment.

A detailed study of the Act has been made by Cooper (1978) and its implications have been considered by Jones (1985) and Mearns and Barzdo (1981).

The animal must be kept in accommodation from which it cannot escape and the premises must be inspected annually by a veterinary surgeon. The licensee must hold insurance, approved by the local authority, against damage caused by the animal

Zoo Licensing Act 1981: any collection of animals not normally domesticated (see Chapter 1 for definition) in Great Britain to which the public has access on more than seven days in any 12 month period.

The terms of the Act are such that many collections not normally described as a zoo may require to be licensed under the Act. It could include a live exhibit (of invertebrates, for example) in a museum, a wildlife rehabilitation centre or a herd of deer which the public are permitted to visit. It is possible to apply for exemption from the Act as a whole or in part from the inspection requirements.

Zoos must meet standards set by the Department of the Environment (DOE) and must undergo a three yearly periodical inspection by inspectors from a list provided by the DOE and appointed by the local authority. Special inspections and informal inspections may be carried out at shorter intervals and for specific purposes. The local authority is also responsible for enforcing health and safety legislation and codes of practice for zoos (HSC, 1985) (see Chapter 8).

The Act has been considered in detail by Cooper (1983) and its implications have been explained by Leeming (1984). See Appendix 3 (Note 12).

CASES

Air India *v*. Wiggins [1980] 2 All E.R. 593, HL(E).
Barnard *v*. Evans [1925] 2 K.B. 794.

British Airways Board *v*. Wiggins [1977] 3 All E.R. 1068.
Chalmers *v*. Diwell [1976] Crim. L.R. 134, D.C.
Ford *v*. Wiley (1889) 23 Q.B.D. 203.
Hudnott *v*. Campbell (1986) *The Times*, 27 June 1986.
McEwan *v*. Roddick [1952] N.Z.L.R. 938.
Rowley *v*. Murphy [1964] 1 All E.R. 50.
Starling *v*. Brooks [1956] Crim. L.R. 480.
Steele *v*. Rogers (1912) 76 J.P. 150.
Swan *v*. Saunders (1881) 45 J.P. 522.

REFERENCES

Anon (1986). Dealing with farm casualties. *Veterinary Record* **119**, 585–586.

Cassell, D. (1987). *The Horse and the Law*. David & Charles, Newton Abbot.

CITES (1980). *Guidelines for Transport and Preparation for Shipment of Live Animals and Plants*. Secretariat of the Convention on International Trade in Endangered Species of Wild Fauna and Flora, Gland.

Collins, N.M. (1987). Legislation and regulations affecting butterfly houses. In *Butterfly Houses in Britain. The Conservation Implications*. International Union for Conservation of Nature and Natural Resources, Gland and Cambridge.

Cooper, J.E., Gibson, L. and Jones, C.G. (1980). The assessment of health in casualty birds of prey intended for release. *Veterinary Record* **106**, 340–341.

Cooper, M.E. (1978). The Dangerous Wild Animals Act 1976. *Veterinary Record* **102**, 457–477.

Cooper, M.E. (1983). The Zoo Licensing Act 1981. *Veterinary Record* **112**, 564–567.

Cooper, M.E. and Cooper, J.E. (1981). The use of animals in films: a veterinary and legal viewpoint. *BKSTS Journal* **63** (9), 544–546.

FAWC (1984). *The Farm Animal Welfare Council: Background Notes on the Council and its Work*. Farm Animal Welfare Council, Tolworth.

FAWC (1985a). *Report on the Welfare of Farmed Deer*. Ministry of Agriculture, Fisheries and Food (Publications), Alnwick.

FAWC (1985b). *Report on the Welfare of Livestock when Slaughtered by Religious Methods*. HMSO, London.

FAWC (1986a). *Egg Production Systems — an Assessment*. Farm Animal Welfare Council, Tolworth.

FAWC (1986b). *Regulations Working Group. Interim Statement*. Farm Animal Welfare Council, Tolworth.

Hansard (1986). Weekly Hansard Issue No. 1398, 24–28 November 1986, Issue No. 1394, 150. HMSO, London.

Hill, J.R.S. (1984). Cruelty reviewed. *Australian Veterinary Journal* **61**, 245–247.

HSC (1985). *Zoos — Safety, Health and Welfare Standards for Employees and Persons at Work*. HMSO, London.

IATA (annual). *IATA Live Animals Regulations*. International Air Transport Association, Geneva.

Jones, M. (1985). Some effects of the Dangerous Wild Animals Act 1976 with regard to the British adder *Vipera berus* in captivity. *British Herpetological Society Bulletin* **11**, 36–37.

Leeming, D. (1984). The Zoo Licensing Act 1981. *British Veterinary Zoological Society Newsletter* **18**, 4–9.

MAFF (undated). *Code of Practice for the Transport by Air of Cattle, Sheep, Pigs, Goats and Horses*. MAFF, Tolworth.
MAFF (various). Codes of Recommendations for cattle, sheep, pigs, domestic fowls and turkeys. Ministry of Agriculture, Fisheries and Food (Publications), Alnwick.
Mearns, C.S. and Barzdo, J. (1981). A survey of the Dangerous Wild Animals Act application. *Ratel* **8**(2), 12–27.
Muriel, K. (1985). Export of live food animals — RSPCA submits formal complaint to European Commission. *RSPCA Today* **48**, 24–25.
Muriel, K. (1986a). Hedgehog case — definition of 'captive' under the Protection of Animals Act, 1911. *RSPCA Today* **53**, 21.
Muriel, K.B. (1986b). *The Welfare of Animals in Transit. Defects in Current Legislation*. Royal Society for the Prevention of Cruelty to Animals, Horsham.
Muriel, K., Mews, A. and Milner, F. (1985). *RSPCA Complaint to the Commission of the European Communities on the International Transport of Live Animals*. Royal Society for the Prevention of Cruelty to Animals, Horsham.
Post Office (annual). *Post Office Guide*. Post Office, London.
Sophian, T.T. (1972). *Horses and the Law*, 2nd edn. J.A. Allen, London.
Weatherill, J. (1979). *Horses and the Law*. Pelham, London.
Wilkins, D.B. (1987). The Welfare of Animals in Films and on the Stage. In *The Welfare of Animals in Captivity*. British Veterinary Association London.
WSPA. (1981). Rating Films from an Animal Welfare Point of View. *Animals International* 1(4), 8–9.

RECOMMENDED READING

See Chapter 1 for literature generally applicable.

Bach, J. (1987). Animals in transit and at markets. In *Legislation Affecting the Veterinary Profession in the United Kingdom*. RCVS, London.
Britt, D.P. (ed.) (1985). *Humane Control of Land Mammals and Birds*. Universities Federation for Animal Welfare, Potters Bar.
Chapman, M.J. (ed.) (1986). *Forensic Aspects of Disease and Husbandry*. M.J. Chapman & Associates, London.
RSPCA (annual). *Annual Report*. Royal Society for the Prevention of Cruelty to Animals, Horsham.
RSPCA (continuing). *RSPCA Today*. Royal Society for the Prevention of Cruelty to Animals, Horsham.
RSPCA (undated). *Cruelty to Animals and the Law*. Royal Society for the Prevention of Cruelty to Animals, Horsham.
Scott, W.N. (1978). United Kingdom legislation concerning farm animal welfare. In *The Care and Management of Farm Animals* (Scott, W.N., ed.). Ballière Tindall, London.
WSPA (continuing) *Animals International*. World Society for the Protection of Animals, London.

4 Animals used for Scientific Purposes

Those using conscious animals should apply to their studies such tenets as Russell and Burch's "3R" principle of reduction, replacement and refinement, Dr Carol Newton's "3S" principle of good science, good sense and good sensibility, and Dr H.C. Rowsell's "3R" tenet: the right animal for the right reasons.

("Guide to the Care and Use of Experimental Animals", Vol. 1, Canadian Council on Animal Care, 1980)

HEADNOTE

This chapter was prepared in the time between the passing of the Animals (Scientific Procedures) Act 1986 on 20 May 1986 and its coming into force on 1 January 1987 (see Appendix 3 (Note 13)). The practical application of the Act was in the course of development at that time and guidance notes, codes of practice and other information were available only in draft form. In addition, it is inherent in the Act that much of the subsequent practice and procedure is open to amendment and improvement in the light of experience. Consequently, it is the responsibility of the reader to ascertain the most recent interpretation of the Act and its implementation.

In view of the phased transition of authorisation between the old and new legislation and since a number of concepts are based on, or carried forward from, practice under the former law, an account of the Cruelty to Animals Act 1876 is included in the Appendix to this chapter.

LEGISLATION ON THE USE OF ANIMALS FOR SCIENTIFIC PURPOSES

The first law relating to animals used for scientific purposes applied to Great Britain and Ireland. This was the Cruelty to Animals Act 1876, which remained on the statute book unchanged (except in its application to Eire) until it was replaced by the Animals (Scientific Procedures) Act 1986. The latter received the Royal Assent on 20 May 1986 and for the most part came into force on 1 January 1987 although a few parts will be brought into operation during the following three years (see Table 1, p. 73) (Richards, 1986).

The new law is part of a worldwide trend towards the development of legislation on this subject. Many countries formulated or revised their laws in the 1970s or early 1980s (see Chapter 9).

The Council of Europe Convention for the Protection of Vertebrate Animals used for Experimental and other Scientific Purposes (Griffin, 1985; Hovell, 1985; COE, 1986a) was adopted by the Council in May 1985, was opened for signature on 18 March 1986 and on 31 March 1986 received the signatures of sufficient member countries, including the UK, to bring it into force on 1 October 1986, prior to ratification by individual countries. The European Economic Community adopted a directive on the subject on 24 November 1986 which requires member countries to comply with its provisions within three years (EEC, 1986). In all respects the new UK law already satisfies the European legislation and in some areas it is stricter.

The new legislation, which applies to Great Britain and Northern Ireland, implements the requirements of the Convention and makes statutory certain powers of the Home Secretary and others which were formerly exercised as a matter of administration and practice.

The Animals (Scientific Procedures) Act 1986 has produced a substantial revision of the regulatory procedures for the use of animals in research in the UK. It will provide for closer scrutiny and supervision, and an increased number of procedures, animals and their users fall within the controls of the new Act. This, in its turn, may well enlarge the number of establishments, licence holders and procedures reported in the annual statistics, although the introduction of the payment of fees may lead to some reduction in animal usage.

The 1986 Act provides for the production or approval of guidance notes and codes of practice by the Home Secretary. The Act requires them to be made public by being laid before Parliament. In particular, the Home Office Guidance on the Operation of the Animals (Scientific Procedures) Act 1986 (HOG) (Home Office, 1986a) provides practical working advice on how the Home Office intends to apply the legislation in practice; it is therefore referred to extensively in the course of this chapter. Further guidance is given in the licence and certificate application forms. Other relevant literature includes:

Guidelines on the Care of Laboratory Animals and their Use for Scientific Purposes (RS/UFAW, 1987);

Guidelines for the Recognition and Assessment of Pain in Animals (Sanford *et al.*, 1986);
Guidelines on the Recognition of Pain, Distress and Discomfort in Experimental Animals and an Hypothesis for Assessment (Morton and Griffiths, 1985);
Guidelines for Veterinary Surgeons employed in Scientific Procedures Establishments and Breeding and Supplying Establishments (BVA/RCVS, 1987)
Guidelines on the Use of Living Animals in Research (Fox, 1986).

Draft guidelines have been produced for duties of "named persons", compiled by the Institute of Animal Technology (IAT, 1987).

Prepared in conjunction with the 1876 Act but applicable also to the new legislation is the Biological Council's (1984) *Guidelines on the Use of Living Animals in Scientific Investigations.*

Some institutions have an ethical, research review or user committee which monitors animal usage and licence applications. While in some countries such bodies play an important part in the official regulation of animal research (see Chapter 9) they have no legal status in UK law (Britt, 1983, 1985a, 1985b; Cooper, 1985; LAEG, 1986; Tuffrey, 1987).

In addition to the Animals (Scientific Procedures) Act, there is other legislation which is relevant to laboratory animals and of which those using or caring for laboratory animals must be aware (Anon, 1987) (see Chapters 3, 5 and 6).

ANIMALS (SCIENTIFIC PROCEDURES) ACT 1986

The Act specifies the animals and types of procedures which it regulates and provides for the authorisation of licensees, of projects and of establishments used for scientific procedures and for the breeding and supplying of those animals. It also deals with the administration of such matters by the Home Secretary, the powers of Home Office inspectors, the establishment of the Animal Procedures Committee and the enforcement of these provisions. The Act is summarised for quick reference in Table 2, p. 74.

Main Principle of the Act

The Act requires that experimental or other scientific procedures using living vertebrate animals must be authorised by both a personal and a project licence and performed at a designated establishment. Section 3 requires that the application of a *regulated procedure* to a *protected animal* must be carried out:

(a) By a person holding a *personal licence* qualifying him to carry out a regulated procedure of that description on an animal of that description
(b) As part of a programme of work specified in a *project licence*, which must authorise the use, as part of that programme, of that kind of regulated procedure with that kind of animal
(c) At a *place* (normally a *designated scientific procedure establishment* (s. 6)) specified in both the personal and the project licence

Protected Animals

A protected animal is defined, by section 1(1), as ''any living vertebrate other than man'' (HOG 3–6). This is supplemented as follows:

(a) Living — an animal is treated as living (and therefore within the provisions of the Act) until there is ''permanent cessation of circulation or the destruction of the brain'' (s. 1(4))
(b) Vertebrate — the term includes:
    (i) ''Any animal of the Sub-phylum Vertebrata of the Phylum Chordata'' (s. 1(5))
    (ii) Immature forms of vertebrates — in the case of mammals, birds and reptiles protection also applies during the second half of their gestation or incubation period; for any other species it applies once they become capable of independent feeding (s. 1(2))

This last provision will bring within the licensing requirements for the first time a number of immature forms and, therefore, procedures; thus the provision applies, for example, to avian eggs after half their incubation time has elapsed, requiring, for example, the CAM (chorio-allantoic membrane) test to be licensed.

The Act makes provision for the Home Secretary to extend the definition of protected animals to include invertebrates of any description or to alter the protection for immature animals (s. 2(3)).

Regulated Procedures

Definition
A regulated procedure is defined by section 2(1) as ''any experimental or

other scientific procedure applied to a protected animal which may have the effect of causing that animal pain, suffering, distress or lasting harm''. For convenience this widely drawn definition will be referred to as the ''specified pain effect'' in any ensuing discussion of the term.

The definition of a regulated procedure encompasses work such as antibody production and the passage of tumours and infective organisms which was not controlled by the previous legislation. Scientific work on free-living animals has become subject to control and the special considerations involved are discussed later (see Regulated procedures involving free-living animals).

The Home Office Guidance indicates that the specified pain effect will be interpreted in the widest terms including ''death, disease, injury, physiological or psychological stress, significant discomfort or any disturbance to normal health whether immediate or in the longer term'' (HOG 6). In assessing the effect which a procedure may have upon an animal, the use of an anaesthetic, analgesic, decerebration or other method of producing insentience must be disregarded (s. 2(4)).

Additional procedures

Section 2(1) also applies to any scientific or experimental procedure which forms part of a series or combination of regulated procedures (whether the same or different) which, when applied to a particular protected animal, may cause the specified pain effect (s. 2(2)). Likewise, procedures leading to birth or hatching with the specified pain effect are also regulated procedures (s. 2(3)), e.g. the breeding of hypertensive or obese rats and nude or athymic mice (HOG 48).

The administration of an anaesthetic, analgesic, decerebration or like procedure which is applied to a protected animal for the purposes of a scientific or experimental procedure also constitutes a regulated procedure (s. 2(4)).

Exceptions

Certain procedures are not classed as regulated procedures and therefore do not attract control under the Act (HOG 6–10):

(a) The ringing, tagging, marking or use of any other humane method of identifying an animal is excluded provided that it causes no more than momentary pain or distress and no lasting harm (s. 2(5))
(b) The administration of a substance or article by way of a medicinal test on animals made under product licences or test certificates in compliance with sections 32(6) and 35(8)(b) of the Medicines Act 1968 (s. 2(6)) is excluded

(c) The killing of a protected animal for experimental or other scientific use in a designated establishment is excluded provided that it is carried out by a method appropriate to that species of animal as specified in Schedule 1 to the Act (s. 2(7)).

The Schedule sets out the methods of euthanasia for particular kinds of animals which consequently do not require specific approval by the Home Secretary. The Schedule lists, for example, the killing of any species of protected animal by way of an overdose of anaesthetic administered by injection (followed by certain procedures to ensure that death ensues). Administration by inhalation is a method of anaesthesia listed in the Schedule in respect of many smaller species. Dislocation of the neck, concussion and the use of carbon dioxide are also permitted for certain species and in certain circumstances. Decapitation followed by destruction of the brain of cold-blooded vertebrates is also included in the Schedule. Although permitted in law, it should be noted that the humane aspects of this procedure are currently under examination by a working party (Cooper *et al.*, 1986; Cooper, 1987).

Any method not specified by Schedule 1, or one listed on the Schedule which is applied to a species other than those for which the method is allowed, which is used in a designated establishment to kill an animal for scientific or experimental purposes constitutes a regulated procedure. It must therefore be authorised by the appropriate personal and project licences. Without authorisation the use of such a method would be an offence under section 3. The Home Office Guidance indicates that methods other than those listed in Schedule 1 will be permitted by the Home Secretary and carried out as regulated procedures authorised by personal and project licences (HOG 9).

Section 2(9) gives the Home Secretary power to alter the Schedule.

Euthanasia does not constitute a regulated procedure when it is performed on protected animals in the course of projects carried out outside designated establishments; for example, in the case of field research or an escaped protected animal, it might be necessary to shoot the animal

(d) Animal procedures which are not experimental and not scientific or which do not, or are not likely to, cause pain, suffering, distress or lasting harm are not controlled by the Act

(e) The Act excludes from the term "scientific procedure" any recognised veterinary, agricultural or animal husbandry practice (s. 2(8)) (see Chapter 6).

Consequently, procedures such as caesarian sections, diagnostic tests or de-horning carried out for veterinary, agricultural or husbandry

> purposes do not constitute regulated procedures. However, if such procedures are performed by novel methods or for experimental purposes which result in the specified pain effect of section 2(1) they fall within the Act and are regulated procedures (HOG 10)

Any *experimental* procedure involving the specified pain effect must be licensed whether or not it is a recognised veterinary practice or performed by a veterinary surgeon. If *any other scientific procedure* having the specified pain effect is carried out as a recognised veterinary practice it must be undertaken either by a veterinary surgeon or under one of the exemptions in the Veterinary Surgeons Act which permits the performance of veterinary treatment by the owner, an employee of the owner, or, by any person, of first aid in an emergency (see Chapter 6).

In any circumstances where it is not clear whether a procedure fails to be regulated by the Act, the advice should be sought of a Home Office inspector. Performing a regulated procedure without the requisite authorisation is an offence under section 3 of the Act; any procedure not covered by the Act is open to prosecution under section 1 of the Protection of Animals Act 1911 or the Protection of Animals (Scotland) Act 1912 if it is shown that the animals concerned suffered unnecessarily.

Personal Licence

Authority

A personal licence confers on a person certain powers in respect of animal research. It indicates the holder's competence and suitability to carry out the work which is authorised by the licence, subject to any limitations or conditions which are imposed therein. It may be used only in conjunction with a project licence (which may be held by the personal licensee or by another person) permitting the performance of the regulated procedures authorised in the personal licence as part of an approved programme of work (HOG 27).

A personal licence is issued by the Home Office and qualifies the holder to (s. 4(1))

(a) carry out a specified regulated procedure
(b) on an animal of specified description
(c) at a specified place

Duration

A licence may be issued for an unspecified duration but it must be reviewed at least every five years. Undergraduate licences are reviewed annually (HOG 38).

A licence may be varied or revoked by the Home Secretary on his own initiative for breach of a condition or for other reasons or at the request of the licensee (s. 11) (see later). A licence must be returned to the Home Secretary for variation or removal of special conditions; it cannot be varied by an inspector except when permission is granted to work temporarily at other premises (HOG 37).

Application

An applicant must be aged 18 years or over (s. 4(4)) and must normally have five O-levels or other acceptable training and experience (HOG 28, 30).

Application for a licence is made on the form in HOG Annex J together with any further information requested by a Home Office inspector (s. 4(2)). The applicant must specify his qualifications, the animals to be used, the procedures and, for a first-time applicant, the projects proposed and any other information requested by the Home Office (HOG 32). This is required to enable the Home Secretary, primarily through a Home Office inspector, to assess the work proposed, the applicant's experience and knowledge, his ability in the care of animals, anaesthesia, analgesia and euthanasia and his general suitability and competence. A specific description must be included for the use of novel procedures, complex surgery or neuromuscular relaxants, which must not be used instead of anaesthesia (s. 17), or micro-surgery (HOG 34).

Sponsorship

An application for a personal licence must be sponsored by another personal licence holder who must be someone of authority at the place where the applicant will use his licence and who must have knowledge of, and certify, the relevant qualifications, training, experience, competence and character of the applicant (s. 4(3), HOG 31). In the case of overseas workers and those for whom English is not their first language the sponsor must also certify their proficiency in English and their understanding of the Act (HOG 30).

Sponsorship is not generally required for those who were licensed under the 1876 Act when applying for a personal licence. However, an application for procedures which are novel or beyond the scope of a licensee's experience may require a sponsor (HOG 29).

Assessment of applicant

The issue of a licence and its scope depend upon the applicant's competence and suitability in the light of his qualifications, training and experience and upon assessment of him by his sponsor and the Home Office inspector (HOG 28) who makes a recommendation on the application to the Home Secretary (HOG 59).

In making his assessment the Home Secretary is obliged to consult a Home

Office inspector and may also consult an independent assessor (s. 9(1)). The Home Secretary must notify the applicant of his intention to consult an assessor and the applicant is entitled to make representations on this matter which the Home Secretary must take into consideration (s. 9(2)). The assessor is appointed from a panel of experts in the biological sciences. If the applicant objects to the proposed assessor an alternative acceptable to both sides will be sought. If this is not possible the applicant must accept the nominated assessor or withdraw his application (HOG 35, 60, 61). The Home Secretary may also refer an application to the Animal Procedures Committee (s. 9(1)) and he will seek its advice when an application relates to cosmetics, tobacco products or microsurgery (HOG 35, 61).

Restrictions

A new or overseas licensee is likely to be subject to supervision for the first year of his personal licence (HOG 30, Annex J). Undergraduate licensees are supervised by a personal licensee named in their licence until they graduate. In other cases the supervisor is not named but is normally a head of department or senior licensee. The personal licensee remains responsible for the welfare of his animals and the project licence holder has the duty to ensure that such licensees comply with the terms of the project licence. The degree of supervision to be exercised will depend upon the competence of the licensee.

A licensee will normally be restricted to a specific project named in his personal licence for the first six months of its duration; when this restriction is lifted he may carry out work which is authorised by his licence under any other project licence. Undergraduate licences are always restricted to named projects (HOG 33).

Responsibilities of a personal licence holder

A personal licensee (or all licensees in a conjoint experiment) is personally responsible (HOG 36) for:

(a) Ensuring that the regulated procedures which he carries out are covered by a project licence and his personal licence

(b) Performing personally the procedures authorised by his personal licence (HOG 13).

He cannot delegate his authority. However, the use of unlicensed assistants in certain routine procedures and for mechanical duties in the course of surgery is permitted in certain circumstances (s. 10(4), HOG 13) which are specified in HOG Annex B.

It is an offence for a project licence holder to procure or permit a person to carry out a regulated procedure which is not authorised by that person's personal licence or by the project licence (s. 22(2)). A person carrying out regulated procedures without, or without the reasonable belief that he has, the authority of a personal or project licence commits an offence under sections 3 and 22(1) (see Table 3, p. 76)

(c) Observing the limitations and conditions of his licence
(d) Seeking advice on his animals from animal technicians, senior colleagues, named person, named veterinary surgeon and inspector
(e) Providing care (by someone able to recognise and deal with pain and distress) for his animals when he is absent and in particular ensuring that another personal licensee or the named veterinary surgeon is available to implement the severity or termination condition
(f) Being present at any time when an animal is most likely to be in pain or distress, especially when recovering from a procedure
(g) Keeping records of animal usage and procedures while this remains the responsibility of the personal licensee (it will eventually be transferred to project licence holders)

Limitations and conditions imposed on a personal licence

Limitations are imposed upon all personal licences and represent the statutory obligations of the licensee. They are set out in HOG Annex C. Failure to comply with the content of the limitations also constitutes an offence under the Act (see Table 3).

Conditions may be imposed by the Home Secretary as he thinks fit (s. 10(1)). Breach of a condition is not an offence and does not affect the validity of a licence but may be cause for revocation or variation (see later) of a licence (s. 10(7), HOG 26). Special conditions may be added to individual licences as requiring, for example, supervision of the licensee or his attachment to one particular project.

The standard conditions can be summarised as follows:

Condition 1	Cages must be clearly labelled and identify the project, procedures and responsible personal licensee
Condition 2	The licensee must take measures to prevent or minimise pain, distress or discomfort in animals used
Condition 3	The licensee must notify the project licence holder if the severity level in condition 1 of the project licence is exceeded or likely to be exceeded
Condition 4	If an animal is in "severe pain or severe distress which cannot be alleviated" the personal licensee must provide euthanasia forthwith by a method

	approved by his licence or in Schedule 1 (see earlier). This is the "inviolable termination condition" required by s. 10(2)(b))
Conditions 5, 6	The personal licensee is responsible for arranging for the care and welfare of his animals in his absence and for obtaining veterinary attention when required
Condition 7	A licensee under supervision must comply with the supervisor's requirements
Condition 8	Home Office permission is required before moving an animal being used for regulated procedures out of a scientific procedure establishment
Conditions 9, 10	If an animal which has been used for a regulated procedure is to be released to the wild, to a farm or as a pet, a certificate of fitness must be obtained from a veterinary surgeon, and if it is to be sent for slaughter for animal or human consumption permission must be obtained from the inspector
Condition 11	If no anaesthesia is to be used, regulated procedures are restricted to those which are no more severe than a simple inoculation or superficial venesection
Condition 12	When anaesthesia is used it must be of sufficient depth to prevent the animal being aware of pain caused, and when a procedure involves recovery from anaesthesia the animal must be cared for according to the standards applied in veterinary practice
Condition 13	The use of neuromuscular blocking agents must be notified to a Home Office inspector 48 hours in advance when the licensee is using them for the first time
Condition 14	The use of a procedure to obtain or demonstrate manual skill requires the authority of a project licence

Project Licences

Authority

A project licence is issued by the Home Secretary to authorise regulated procedures to be performed on animals by persons who hold personal licences. A project licence specifies "a programme of work" and authorises "the application, as part of that programme, of specified regulated procedures to animals of specified descriptions at a specified place or specified places" (s. 5(1)). Programmes of work can be of very varied size, scope, duration and animal usage, ranging from a simple research or teaching study of a few animals to fundamental research development of a new surgical technique or to long-term toxicity testing (HOG 39). Licences will be matched to the competence and facilities available to the applicant (HOG 40, 41).

Project licences are granted only for programmes satisfying one or more specified purposes (s. 5(3)) as follows:

(a) The prevention, diagnosis or treatment of disease, ill-health or abnormality in man, animals or plants
(b) The study or modification of physiological conditions in man, animals or plants
(c) Environmental protection for the benefit of human or animal health and welfare
(d) Research in the biological and behavioural sciences
(e) Education or training (regulated procedures are not permitted in primary or secondary schools, however)
(f) Forensic enquiries
(g) The breeding of animals to be used for experimental or other scientific purposes

Duration

The duration of each licence is specified by the Home Secretary and is renewable with an overall maximum time of five years (s. 5(7)); if the project runs for longer, a further licence will be required. The Act provides for limited continuity of the licence on the death of the holder (s. 5(8)). An inspector must be informed if the holder ceases to fulfil the legal or licence requirements, as, for example, on a change of employment or retirement (HOG 67). A licence may be varied or revoked by the Home Secretary or at the instigation of the licence holder (s. 11) (see later).

Application

Application for a licence is made on the form in HOG Annex I. It will include the applicant's proposed programme of work which should include the procedures, their severity and expected benefit, animal usage, pain control and euthanasia, and special consents required for re-use or neuromuscular blocking agents (HOG 40).

Assessment of licence application

In assessing a project the Home Secretary will consider (HOG 58):

(a) The project licence applicant's status and qualification
(b) Animal usage in terms of refinement, replacement and reduction, e.g. whether the proposed work could be achieved otherwise than by using protected animals (s. 5(5)) or by using fewer animals or milder procedures and whether the programme, species and methods of pain control and the facilities available for the care and housing of the animals to be used are suitable
(c) The justification for the use, if requested, of:
  (i) Cats, dogs, primates or Equidae on the grounds that no other species are suitable or can practicably be obtained (s. 5(6))
  (ii) Cats and dogs which have not been bred at and obtained from a designated breeding establishment for the reason solely that no suitable cat or dog can be obtained (s. 10(3)(a))
  (iii) Other Schedule 2 species which have not been bred at a designated breeding establishment or obtained from a designated supplying establishment (s. 10(3)(b))
(d) The cost/benefit (s. 10(5)) and the severity of the project (see later)

The Home Secretary may refer an application (particularly in respect of new or highly specialised work) to an independent assessor chosen from a panel of experts in the biological sciences following the procedure outlined earlier in the case of personal licences. He may also refer a proposed project in a particular field, such as cosmetics or microsurgical training, to the Animal Procedures Committee (s. 9(1), HOG 60, 61).

Cost/benefit assessment

Section 5(4) of the Act requires the Home Secretary to make a cost/benefit assessment by weighing "the likely adverse effects on the animals concerned against the benefit likely to accrue as a result of the programme to be specified in the licence". This will be a matter of weighing the severity of a procedure against the importance of the purpose of the project.

Severity

In making this assessment the Home Secretary pays particular attention to the need to control and monitor the levels of severity in regulated procedures.

Severity is interpreted in the widest terms, to include the "specified pain effect" of section 2(1) (see earlier) as well as, in the words of the Guidance, "impairment of health or well-being, morbidity, mortality and any other factor which may affect the animals used in procedures" (HOG 51).

Factors which may affect the level of severity will include the effects of surgery, environment, toxicity, infection, physiological stress and the duration of a procedure; severity may be reduced on account of the anaesthesia, analgesia, sedatives, nursing and general care which are provided. When terminally anaesthetised animals or decerebrate animals are used some of these factors may not be relevant (HOG 51). In all cases project licence condition 6 and personal licence condition 2 (s. 10(2)(a)) require that licensees adopt, and maintain with pain-reducing measures, the lowest possible level of severity which is compatible with achieving the purpose of any procedure.

The levels of severity are divided into bands which are defined by the Guidance as "mild", "moderate" or "substantial" (HOG 52). Procedures carried out under anaesthesia without recovery or on decerebrated or otherwise insentient animals are described as "unclassified" (HOG 50, Annex I).

"Mild" severity includes the taking of a blood sample or dosage with substances which is little different from normal physiological limits, simple skin tests anticipating mild reactions, most antiserum production, laparoscopy or biopsy under anaesthesia with recovery (HOG 52).

The "moderate" level of severity includes many pharmaceutical screening and toxicity tests (with certain limitations or provisions for euthanasia), studies of disease when the effects are moderate or when death is unlikely to be accompanied by suffering and surgical procedures with recovery from anaesthesia (HOG 53).

The "substantial" category takes in the most severe procedures including most lethality testing (HOG 54).

There is also an overall limit for all regulated procedures known as the "inviolable termination condition" (personal licence condition 4, s. 10(2)(b)) which will not only operate in unforeseen circumstances but also to prevent procedures of extreme severity being undertaken or authorised.

While many procedures fall clearly into one of the three categories others may not or there may be some overlapping. In all cases the level depends on the pain it may cause and not on the procedure involved. Hence, it can be said that while a particular procedure normally falls into a certain category, the method used, the frequency (e.g. of blood sampling) or quantity may alter the category of severity. Likewise, the duration of the procedure or its effects are factors in the severity assessment (HOG 52, 55).

An applicant for a project licence is required to make an assessment of the severity band appropriate to the proposed regulated procedures (HOG 49,

52) and to the need to minimise severity in accordance with condition 6 (HOG 55). The setting of the level is discussed and negotiated between the Home Office and the licence applicant and may lead to modification of a project proposal. The overall banding which is finally agreed (not necessarily the same for all procedures which are authorised) is inserted in condition 1 of the project licence but may be revised subsequently (HOG 55–57).

Exceeding a severity band

A personal licensee is responsible for the welfare of the animals which he uses and for ensuring that condition 12 of his licence regarding anaesthesia is observed. He must notify the project licensee if he anticipates that a severity band is likely to be exceeded (HOG 36, personal licence condition 3).

A project licence holder who foresees an excess of severity should either stop the procedure or consult the inspector. The latter will then either terminate the procedure or temporarily (i.e. for up to 14 days, subject to the Home Secretary's confirmation) permit a higher level of severity. Failure to report a foreseeable rise in severity is a breach of project licence condition 1 and therefore grounds for revocation or variation of a licence. An unpredictable rise in severity does not constitute a breach of the condition (HOG 62–64).

When a severity band appears to have been exceeded (or if it is likely to be exceeded) the project licence holder must notify the Home Office inspector promptly (condition 1) and give attention to reducing the suffering of the animal concerned (HOG 63).

Obligatory euthanasia

The Act and Guidance also provide that an animal which suffers in the course of a regulated procedure must be killed humanely in the following circumstances (see Euthanasia, page 66):

(a) At the end of a regulated procedure (or series thereof) an animal which is "suffering or likely to continue to suffer adverse effects" (s. 15(1), personal licence limitation 9, HOG 17)
(b) At any time when an animal is in "severe pain or severe distress which cannot be alleviated" — this is the "inviolable termination condition" (personal licence condition 4, s. 10(2)(b))
(c) When required by a named veterinary surgeon or named person because he is concerned for its health or welfare (s. 66(b), s. 7(6))
(d) When required by a Home Office inspector because an animal is suffering excessively (s. 18(3), personal licence limitation 10, HOG 20)

In view of the responsibility of both project and personal licensees for the recognition of suffering, pain, discomfort, distress in the animals which they use, reference should be made to current publications on the subject, such as Morton and Griffiths (1985) and Sanford *et al.* (1986) (HOG 65).

Responsibilities of a project licence holder

A project licence is granted to the person who takes overall responsibility for the work authorised by it (s. 5(2)). His responsibilities can be summarised as follows.

The project licence holder:

(a) Must normally be a personal licence holder: if not, a deputy who is a personal licence holder must also be appointed.

A deputy may also be required to cover the absence of the main licence holder or to oversee a particular location or part of a project (HOG 43–45)

(b) Must be a senior person of standing and authority at the place where the project licence work is to be performed

(c) Has overall responsibility for the design and conduct of the project (s. 5(2), HOG 43) and for compliance (by himself and other personal licensees) with the Act and the conditions in the project licence (HOG 42)

(d) Must ensure that personal licensees are duly authorised and know the severity limit imposed on the project (HOG 42)

(e) Must notify the inspector if a severity limit has been or is likely to be exceeded (condition 1)

(f) Must keep records of animals and procedures used and by which personal licensees; make an annual report to the Home Secretary of all usage for that year; provide other reports required by the Home Secretary and any publications relating to work under the licence (conditions 3–5)

Limitations and conditions imposed on a project licence

A project licence is issued subject to limitations and conditions set out in HOG Annex D.

The limitations are drawn up to implement the relevant provisions of the Act. Failure to comply with certain of them is also a breach of statutory requirements and therefore an offence (see Table 3, p. 76).

The conditions in a licence are imposed by the Home Secretary as he thinks fit (s. 10(1)). A breach of condition is not an offence against the Act (HOG 26) and does not affect the validity of the licence; however, it may lead to revocation or variation see later of a licence (s. 10(7)).

Special conditions may be added to individual licences in respect, for example, of duration or severity of procedures and animal usage (HOG 62).

The standard conditions can be summarised as follows:

Condition 1	The project licence holder has a duty to ensure notification of the inspector if a severity limit specified in this condition is (or is expected to be) exceeded
Condition 2 (not immediately in force)	In accordance with section 10(3) and Schedule 2 of the Act, unless authorised by the Home Secretary, the use of mice, rats, guinea pigs, hamsters, rabbits and primates is restricted to those bred at a designated breeding establishment or supplied by a designated supplying establishment; the use of cats and dogs is restricted to those bred and supplied by a designated breeding establishment (note HOG 58 refers to "the designated breeding establishment at which they were bred")
Conditions 3–5	The licence holder is responsible for the maintenance of records, which must be available to the inspector at any time, of the animals used, of the regulated procedures performed and of the licensees working under the project licence, for the submission of the annual report to the Home Secretary containing statistics of usage under the project licence (condition 4 not immediately in force), and for the provision of any other reports required and a list of publications relating to the project work
Condition 6	The level of severity in a procedure must be the lowest consistent with achieving its aim
Condition 7	An animal used under the project licence may not be used for a procedure under a different project licence without specific authority. The project licence holder must ensure that there is no unauthorised re-use of animals under his licence

Re-use of Protected Animals

The Act permits the re-use of protected animals for more than one purpose only in the most restricted circumstances (s. 14, HOG 16).

An animal must not be used for regulated procedures under more than one project licence concurrently, unless otherwise authorised (project licence condition 7).

When a protected animal has been used in a series of regulated procedures for a particular purpose without the use of general anaesthesia at any stage it may not be re-used for further regulated procedures without the consent of the Home Secretary (s. 14(3)).

When, in a series of regulated procedures, general anaesthesia has been given and recovery permitted, an animal may not normally be re-used (s. 14(1); however, with the Home Secretary's consent (in accordance with section 14(2) and personal licence limitation 11) re-use is allowed but only when:

(a) The earlier procedure under anaesthesia was used for surgical preparation essential for a subsequent procedure, or
(b) The anaesthesia was used solely for immobilisation, or
(c) The subsequent procedure is to be performed under general anaesthesia without recovery

The terms on which an animal may be re-used are given in HOG Annex H and can be summarised as follows:

(a) The only reason for permitting re-use is the reduction in use of animals
(b) Re-use must be under terminal anaesthesia, without preparatory procedures
(c) The animal must not be suffering, or likely to be suffering, adverse effects after the first series of regulated procedures (in such a case section 15(1) requires its euthanasia)
(d) A veterinary surgeon must have certified that it is in good health and has not suffered any lasting harm
(e) It must not have been kept for an excessive time before re-use
(f) It may not be moved to another scientific procedure establishment for re-use except in exceptional circumstances
(g) It may not be sent to a supplying establishment for redistribution or resale

An animal which has been used for regulated procedures and which is to be returned to the wild, sent to a farm or be used as a pet must first have been certified as fit by a veterinary surgeon (personal licence condition 9).

Regulated Procedures in Training and Education

The use of regulated procedures in the course of education or training is one of the purposes for which the issue of a project licence is permitted by section 5(3)(e) of the 1986 Act (HOG 46).

Project licences may not be issued for the performance of regulated procedures in primary and secondary education (s. 5(3)(e), HOG 47).

Personal and project licences are issued for undergraduate and postgraduate work with various restrictions. The personal licensees must work under supervision and the project licence for their programme of work must be held by the supervisor or head of department (HOG 46 (ii)).

The demonstration of regulated procedures to illustrate lectures in higher scientific education, in training courses at scientific procedure establishments or at meetings of learned societies may be authorised by a project licence provided that there is no suitable alternative method available. HOG 46 (iii) provides that the use of animals to develop manual skills is permitted only for training in microsurgery using terminally anaesthetised rodents. Applications for project licences for such work are referred to the Animal Procedures Committee. This provision also restricts the training which can be given to prospective licensees in elementary techniques.

The performance of regulated procedures specifically to make a film or video or for teaching purposes is permitted and requires a project licence. However, no further authorisation is required to film on-going regulated procedures under an existing programme of work already authorised by a project licence. Only the filming of a regulated procedure for live transmission for general public viewing is illegal, as is the advertising thereof (see Table 3) (s. 16, HOG 18).

A regulated procedure may not be performed (or advertised) for exhibition to the general public (s. 16). Nevertheless, those with a *bona fide* interest may be permitted to view regulated procedures which are part of an on-going programme (HOG 18).

Designated Establishments

Premises used to carry out regulated procedures (s. 6) and for breeding or supplying protected animals (s. 7) are called "designated establishments" in the Act. They must be designated by a certificate issued by the Home Secretary. This will specify a "named person" responsible for the day-to-day care of animals kept in that establishment and a "named veterinary surgeon" or occasionally (see later) a "suitably qualified person" to provide advice on their health and welfare.

Certificates, issued under sections 6 or 7, are applied for on the form in HOG Annex K (s. 6(3), s. 7(3)) and must nominate a named person and named veterinary surgeon (s. 6(4)(b), s. 7(4)) and continue until they are suspended or revoked. They do not require renewal (s. 6(8), s. 7(8), HOG 75) and may be varied at the request of the holder (HOG 74).

Fees will be charged in respect of certificates issued under section 6 and section 7 in accordance with section 8 of the Act and will consist of an annual flat rate and fee based on the number of personal licensees having primary availability at the establishment at any time in the preceding calendar year (HOG 73). See Appendix 3 (Note 14).

Scientific Procedure Establishments

Authority

Regulated procedures may only be carried out at a place specified in the personal and project licences authorising these procedures (s. 3(c)). Normally such a place must be designated by a certificate of the Home Secretary as a "scientific procedure establishment" (s. 6(1)) although he has the power to specify another place (s. 6(2)). This exception may be used to authorise fieldwork involving non-domesticated animals or occasional research carried out at a farm (HOG 68).

Nominated persons

The certificate of designation must be issued to a person "occupying a position of authority at the establishment in question" (s. 6(4)(a)). A certificate holder is normally a senior person who is in a position to implement changes and obtain finance and who may have been the "responsible authority" under the 1876 Act. He will hold the final responsibility for ensuring that the conditions imposed in the certificate are observed (HOG 68).

The certificate must also specify a person (often referred to as the "named person") responsible for the day-to-day care of the protected animals kept at the establishment for experimental or other scientific purposes (s. 6(5)(a)); it must also specify a veterinary surgeon (the "named (or nominated) veterinary surgeon") who provides advice on the health and welfare of those animals (s. 6(5)(b)). With the approval of the Home Secretary, one person may be designated as both (s. 6(5)); however, in some organisations there may be a named person and veterinary surgeon for separate departments or buildings (HOG 69). Occasionally, a person other than a veterinary surgeon may be named under section 6(5)(b) when there is no local veterinary surgeon with sufficient expertise in a particular species, particularly fish. The

Guidance provides that the Royal College of Veterinary Surgeons will be consulted as to the availability of an appropriately experienced veterinary surgeon before the appointment of a "suitably qualified person" (HOG 70).

Named person and named veterinary surgeon

The Act requires the named person or the named veterinary surgeon who becomes concerned about the health or welfare of an animal to notify the personal licensee having charge of the animal; if this is not practicable (either because the licensee is not available or there is none), the named person or the named veterinary surgeon must ensure it is cared for or, if necessary, killed by a method appropriate under Schedule 1 or approved by the Home Secretary (s. 6(6)). Notification of an animal causing concern may also be made by the named person to the named veterinary surgeon and both have the power to notify a Home Office inspector (s. 6(7), HOG 72).

The general duties and obligations of a named veterinary surgeon are set out in Guidelines on the subject drawn up by BVA/RCVS (1987) and, more briefly, by BVA/RCVS (1986).

A named veterinary surgeon must:

(a) Provide, with the help of deputies, a 24 hour clinical emergency and specialist service.

This will include providing advice on animal care and welfare, animal husbandry, disease control, a diagnostic and quality control programme; on the recognition, assessment and control of pain; on surgical and experimental techniques

(b) Collaborate with the named person, animal technicians, research workers, certificate holder and Home Office inspector

(c) Visit the establishment regularly, as appropriate to its size and operation

(d) Implement his duties, under sections 6 and 7, of notification and euthanasia

(e) Keep records of his advice.

He is also required to keep a health record of the animals at the establishment (Home Office, 1986b)

Other duties of the named veterinary surgeon may include involvement in regulated procedures, laboratory animal husbandry and personnel management, implementation of animal and human health legislation and participation in research and teaching.

Conditions imposed on a certificate of designation

A certificate will be issued subject to standard conditions (HOG Annex E) together with any special conditions that the Home Secretary thinks fit

(s. 10(1)). Breach of a condition is not an offence and does not invalidate the certificate but may lead to revocation or variation (see later) of a licence (s. 10(7), HOG 26).

The standard conditions may be summarised as follows:

Conditions 1–3	The Home Secretary's approval must be obtained for alterations to the establishment, to the use of the designated rooms there, and to the kinds of animals kept at the establishment
Condition 4	The establishment must be appropriately staffed for the care of protected animals
Condition 5	The named person must provide for the prompt euthanasia of any protected animal which is found to be in severe pain or severe distress and for which there is no responsible personal licensee
Condition 6 (not immediately in force)	Section 10(3) must be observed in obtaining animals (see project licence condition 2 and HOG 71) although an exemption may be provided when animals are not available (HOG 71)
Condition 7	Records must be kept of the source, use and final disposal of all protected animals kept at the establishment although section 10(6)(b) refers to animals kept for experimental or other scientific purposes. The records must be available for examination by the Home Office inspector or for submission to the Home Secretary
Condition 8	Primates, dogs and cats used or intended for use in procedures must be marked by a method approved by the Home Secretary
Condition 9	Alterations in the name of the establishment, the certificate holder or the named veterinary surgeon and any change in the name or qualifications of the named person must be notified to the Home Office
Condition 10	The notification procedures in section 6(6) (see earlier) must be observed (HOG 72)
Conditions 11–14	The accommodation and care provided for protected animals must be appropriate; environmental conditions must be checked daily; the escape of animals or intrusions must be prevented; if

	required, quarantine and acclimatisation facilities should be provided
Condition 15	Under the supervision of the named veterinary surgeon records of the health of protected animals must be kept and be available to the inspector
Condition 16	An inspector must have access at all reasonable times to the areas in which protected animals are kept
Condition 17	Adequate fire precautions must be maintained
Condition 18	The certificate holder must take reasonable steps to ensure that unauthorised procedures do not take place at the establishment
Condition 19	The provision in sections 10(5) and 10(6)(a) for euthanasia must be observed (see Euthanasia, page 66)

Breeding and Supplying Establishments

The protected animals listed in Schedule 2 which are intended for use in regulated procedures must be bred at premises designated as a "breeding establishment" by a Home Office certificate (s. 7(1), HOG 76, Annex F).

Schedule 2 species are:

Mouse	Rabbit	
Rat	Dog*	*Must be bred *and* supplied by a designated
Guinea-pig	Cat*	breeding establishment (s. 10(3))
Hamster	Primate	

Any premises where these animals are kept (but not bred) to be supplied for use elsewhere for regulated procedures must be designated by a Home Office certificate as a "supplying establishment" (s. 7(2), HOG 77).

Establishments which qualify for more than one of the certificates must apply for each designation which is appropriate. An establishment which breeds animals with harmful genetic mutations or supplies surgically prepared animals must also be designated as a scientific procedure establishment (HOG 78).

An overseas breeding or supplying establishment cannot be certified under the Act. Consequently, consent is required for the use of all imported cats and dogs and also for the use of imported Schedule 2 species unless they have been acquired from a supplying establishment. While there is no restriction in the Act to importation as such, it would be prudent to seek Home Office approval to ensure that animals may be used once they have been imported.

Provisions comparable to those for a scientific procedure establishment regarding a named person and a named veterinary surgeon (ss. 7(5)–7(7)), the standard conditions (HOG Annex G) attached to a certificate, fees (5.8) and duration of a licence (s. 7(8)) apply to breeding and supplying establishments (HOG 76–79).

Variation, Revocation and Suspension of Licences and Certificates

The Act gives the Home Secretary power to refuse, vary, revoke or suspend a licence or certificate (ss. 11–13, HOG 80–84). Variation and revocation may also occur at the instigation of the holder (s. 11(c), HOG 66).

The Home Secretary may vary a licence or certificate (s. 11) when:

(a) A condition therein has been broken (s. 10(7), HOG 26)
(b) He considers it appropriate
(c) Requested by a licence or certificate holder

When the Home Secretary proposes to refuse, revoke or vary a licence or certificate he must notify the holder (personally or by post (s. 12(8)) of the reasons for doing so and of the holder's right to make oral or written representation (s. 12(1)–12(3), HOG 83) in accordance with rules made under section 12(7). Representation may also be made in objection to a condition imposed in a licence or certificate, although the condition will remain in force until the Home Secretary has made a decision (s. 12(4)), HOG 83).

A legally qualified person will be appointed to hear the representations together with, if the Home Secretary thinks fit, a person with scientific or other appropriate experience (s. 12(5), HOG 84). The assessor makes a report to the Home Secretary who must supply a copy to the person making the representation. The Home Secretary must take the report into account when deciding upon the condition or the refusal, revocation or variation (s. 12(6), HOG 83).

The Home Secretary may suspend a licence or certificate with immediate effect for a maximum of three months if he considers it "urgently necessary for the welfare of any protected animals" (s. 13(1)) and the holder must be notified (personally or by post (s. 13(8)). Further suspensions for periods of up to three months are permitted if negotiations for revocation or variation are continuing (s. 13(2), HOG 81).

On suspension or revocation the Home Secretary will make arrangements for the care of any animals affected by it and to avoid wastage of animals (HOG 82).

Euthanasia

With a view to limiting pain and suffering and ensuring the use of humane methods of killing, the Act makes a number of provisions for the euthanasia of animals kept or used for regulated procedures.

Certain methods are prescribed by Schedule 1 to the Act and HOG Annex A as appropriate for specified kinds and sizes of animals (see earlier). The use of any other method for the killing of a protected animal for experimental or other scientific purposes at a designated establishment itself constitutes a regulated procedure and must be authorised by a project and personal licence (s. 2(7)). If not so authorised, its use constitutes an offence under section 3, even in the event of an emergency.

This restriction on methods of euthanasia does not apply outside designated establishments, when, for example, it may be necessary to use a firearm to kill a wild or escaped animal or to send a farm animal to a slaughterhouse.

The Act and Guidance also provide a number of circumstances in which a protected animal must be put down by an appropriate Schedule 1 method or, alternatively, by another method permitted in the personal licence of the person using the animal for a regulated procedure. These are:

(a) At the end of a series of regulated procedures if the animal is suffering or likely to suffer adverse effects (s. 15(1), HOG 17)
(b) On the conclusion of a regulated procedure in accordance with a project licence condition (e.g. under (c) below) (s. 15(2))
(c) Under the "inviolable termination condition" (s. 10(2)(b)) if at any time an animal is in "severe pain or severe distress which cannot be alleviated" (personal licence condition 4).

    A personal licensee's competence in euthanasia may be part of the assessment of his suitability to hold a personal licence (HOG 34). He is personally responsible for ensuring that the inviolable termination condition can be satisfied in his absence (HOG 36).
(d) The named person and named veterinary surgeon have a duty to provide euthanasia in accordance with Schedule 1 methods or by methods approved by the Secretary of State when acting under section 6(6)(b) and section 7(b) if no personal licensee is available to do so (HOG 78)
(e) When an inspector considers euthanasia necessary because of excessive suffering in a protected animal (s. 18(3))
(f) Under the requirement for designated establishments to use Schedule 1 methods or a method approved by the Home Secretary when killing animals intended, but not actually used, for regulated procedures or

for breeding or supplying such animals and when killing animals not within (a) above and ex-breeding and supplying stock (s. 10(5)).

In this case, the establishment must have available a person competent in the use of such methods (s. 10(6)(a))

These provisions are reinforced by the standard limitations and conditions imposed on licences and certificates (HOG Annexes C–E). In addition, a special condition in a project licence could require the killing of an animal in specified circumstances (HOG 62). It is an offence to fail to comply with paragraphs (a) and (d) above (see Table 3, p. 76).

Records

Records must be kept in respect of regulated procedures and protected animals as required by section 10(6)(b) and by conditions in project licences and in certificates (see earlier).

Guidance has been produced on the obligations of licence and certificate holders (Home Office, 1986b) and instructions on the completion of the licence holder's annual return to the Home Office are contained in the Home Office explanatory notes (1986c).

The records which must be maintained can be summarised as follows:

(a) A personal licence holder must ensure that all cages are labelled to identify the animals, the project and the licensee using them
(b) A project licence holder and deputy must keep detailed records of the animals used, the regulated procedures carried out and the personal licensees working under his licence
(c) Certificate conditions require holders to ensure that cats, dogs and primates in designated establishments are individually and permanently marked for identity
(d) The named veterinary surgeon must maintain health records of protected animals kept at designated establishments
(e) In scientific procedure establishments certain larger species as well as those in (c) above must be readily identifiable. Cages must be labelled, source and disposal records kept and a health record must be maintained under the supervision of the named veterinary surgeon
(f) Special provisions are made for the records relating to immature forms, harmful genetic mutants and wild animals

It is the responsibility of certificate holders to ensure that the information is properly maintained. All records must be kept for five years after the death or disposal of the animal recorded.

Annual Return

An annual return must be made to the Home Secretary in accordance with Home Office (1986c). It gives details of the regulated procedures commenced in that year, of the animals used and of the purposes of the procedures. The annual return is made by personal licensees for the years 1987 and 1988 whereafter it becomes the responsibility of project holders.

The annual returns provide the data from which the Home Secretary annually provides information to Parliament under s. 21(7) and are the basis of the statistics which will be published by the Council of Europe (COE, 1986c).

Public Information: Guidance, Codes and Annual Statistics

In the past information on the Home Office practice for implementing the 1876 Act was mainly provided from outside, albeit in some cases with Home Office approval and collaboration (RDS, 1974, 1979). It has also been described in Home Office (1981, 1983) and parliamentary (Littlewood, 1965; Anon, 1980) reports and by other authors, most recently O'Donoghue (1980), Cooper (1981), Blackman (1985) and Uvarov (1985).

Codes of practice for the care and use of laboratory animals have been produced on a national (Biological Council, 1984) or international (Zimmermann, 1984; CIOMS, 1985; COE, 1986b) basis and individual institutions commonly have local rules for use in their own animal facilities. A number of codes or guidance notes has been prepared with a view to satisfying the requirements of the 1986 Act.

Under the new Act the Home Secretary has a statutory obligation to:

(a) Publish information on the granting of licences and certificates under the Act and on the conditions to which they are subject (s. 21(1)).

This comprises the *Home Office Guidance on the Operation of the Animals (Scientific Procedures) Act 1986* (Home Office, 1986a). Further information is given in the notes to the licence and certificate application forms. In addition, there is information on record keeping (Home Office, 1986b) and on licensees' statistical returns (Home Office, 1986c) and the transitional arrangements for the new legislation (Home Office, 1986d)

(b) Issue codes of practice for the care of protected animals and their use for regulated purposes.

He may approve codes which are issued other than by himself (s. 21(2)).

(c) Consult the Animal Procedures Committee before issuing, altering or approving any of the foregoing material (s. 21(3))

Information and codes of practice issued or approved under section 21 by the Home Secretary and any amendment thereof must be laid before Parliament. Material so laid must be withdrawn if required by a resolution of either House of Parliament within 40 days of its being laid. In such a case, a substitute must be provided (s. 21(5)(b)). This provision ensures that information is generally available and provides for some public influence on its content.

Codes of practice (but not guidance) are given a limited legal status in that, while a failure to comply therewith will not of itself make a person liable to legal proceedings (criminal or civil), they are admissible as evidence and, if considered by the court to be relevant to some issue, they must be taken into account in deciding that matter (s. 21(4)).

The Home Secretary, by order of Parliament, has made an annual report to it on animal experiments since 1877 (Littlewood, 1965). In recent years this report has contained statistical analysis of animal usage by way of the numbers and species of animals used, the procedures performed, the purposes involved and the numbers of licence holders and institutions.

The 1986 Act has imposed a statutory duty upon the Home Secretary to publish his information annually and to lay it before Parliament. The Act gives him a discretion to include any information which he considers appropriate relating to the usage of protected animals for experimental or other scientific purposes (s. 21(7)).

Inspectorate

Home Office inspectors are appointed by the Home Secretary and hold a medical or veterinary qualification (s. 18(1)).

An inspector's statutory duties are to:

(a) Advise the Home Secretary on licence and certificate applications, variation or revocation and on the periodical review of licences (s. 18(2)(a),(b))
(b) Visit places (normally scientific procedure establishments) where regulated procedures are carried out to determine whether the procedures are duly authorised and licence conditions are observed (s. 18(2)(c))
(c) Visit designated establishments to determine whether certificate conditions are being observed (s. 18(2)(d))
(d) Report to the Home Secretary any failure to comply with the Act or with licence or certificate conditions and to advise him on further action (s. 18(2)(e))

An inspector has no right of entry to the premises referred to in (b) and (c) although a refusal of admission could lead to variation, revocation or suspension of the relevant certificate.

A police constable may obtain a warrant to enter any place (using reasonable force if necessary) to search it and to require any person there to give his name and address (s. 25(1)). If the warrant relates to a designated establishment it must require the constable to be accompanied by a Home Office inspector; in the case of other places, this is optional (s. 25(2)).

The provisions of section 24 regarding information received in confidence (see later) is applicable to inspectors (HOG 14).

The Act gives an inspector the power to require the immediate euthanasia of any protected animal which is "undergoing excessive suffering"; this must be carried out by a method authorised by Schedule 2 or by the personal licence of the person required to kill the animal (s. 18(3)). Failure to fulfil such a requirement is an offence (s. 22(3)(b), HOG 20).

In the course of his duties an inspector also provides advice and guidance to licensees, certificate holders, named persons and named veterinary surgeons on the day-to-day application of the Act and Home Office practice.

Animal Procedures Committee

The Animal Procedures Committee is the most recent in a succession of committees charged with advising the Home Secretary on the use of animals in scientific research (Anon, 1986; Balls, 1986). It replaces the Advisory Committee on Animal Experiments and although its scope and constitution are comparable, the new body is the first to have statutory authority (s. 19(1)).

The Animal Procedures Committee comprises a chairman and at least 12 other members of whom one must be a lawyer and at least two-thirds must have a veterinary, medical or biological qualification, the last being in a subject approved by the Home Secretary as relevant to the work of the Committee. At least half the members should not have held a licence under the 1876 or 1986 Acts for at least six years prior to their appointment. There is provision for the resignation or, in certain circumstances, the dismissal of a member. The Home Secretary may remunerate the chairman, reimburse the expenses of members and pay other costs of the Committee. The Act requires the Committee to advise the Home Secretary on matters relating to the Act or his functions under it which are either referred by him or initiated by the Committee itself (s. 20(1)). In so doing it may promote relevant research and seek the advice of other persons with appropriate knowledge or experience (s. 20(4)). As a matter of policy, personal or project licence applications for certain procedures such as the testing of cosmetics or training in microsurgery are referred to the Committee (HOG 61).

In the course of its deliberations the Committee must give due attention "both to the legitimate requirements of science and industry and to the protection of animals against avoidable suffering and unnecessary use in scientific procedures" (s. 20(2)).

An annual report of its activities must be delivered to the Home Secretary who must lay copies before Parliament (s. 20(5)).

The provisions of section 24 regarding the treatment of material received in confidence in the course of exercising functions under the Act apply to members of the Committee and its subcommittees.

Confidentiality

Confidential material such as that which is personal or of commercial or scientific value may have to be disclosed in the course of compliance with the Act, particularly when applying for a licence or certificate. The Act provides that it is an offence for a person who knowingly receives confidential information in the exercise of his functions under the Act to disclose such information otherwise than in the course of these functions (s. 24(1), HOG 14).

Only certain information and circumstances are restricted by this provision. Inspectors or members of the Animal Procedures Committee may communicate confidential material amongst themselves, to assessors or to the Home Secretary. Non-confidential material may be passed on without committing an offence.

The offence relates only to disclosures which are made with the knowledge that, or with reasonable grounds for believing that, the information was given in confidence (s. 24(1), HOG 14). Care must therefore be taken by those who provide sensitive information in the first place to indicate that they wish it to be treated in confidence.

Regulated Procedures Involving Free-living Animals

Regulated procedures performed upon non-domesticated vertebrate animals are subject to the 1986 Act in the same way as those on more conventional species. However, special considerations apply to animals living free in the wild.

Experimental work was not normally permitted under the Cruelty to Animals Act 1876 owing to the difficulty of supervising a free-living subject. The wider definition of a regulated procedure in the 1986 Act has brought under Home Office control a number of "other scientific procedures" which are commonly used in biological fieldwork. These include toe clipping (for identification), blood sampling, the administration of drugs, the use of an immobilising drug, even if solely for the purpose of restraint during procedures not requiring a licence, implanted telemetry devices or other invasions of the body (see also Appendix 3 (Note 15)). Licences are unlikely to be made available to amateur naturalists.

Certain procedures are specifically exempted by the Act, i.e. ringing and other harmless identification methods such as fur clipping (s. 2(5)) and recognised veterinary, agricultural or animal husbandry practices (see earlier) (s. 2(8)). Non-invasive normally harmless work such as taking measurements and attaching radio collars is unlikely to require a licence but in cases of doubt the Home Office should be consulted.

Considerations which are relevant to such work include the following:

(a) If regulated procedures are to take place outside a scientific procedure establishment by way of exception to s. 6(1) the place where the procedure is to be performed must be specified in the relevant personal and project licences (s. 6(2), s. 3(c))
(b) All the provisions relating to licences will be applicable including those for the named veterinary surgeon who may be required to attend regulated procedures which take place in the field
(c) The Act's restrictions on killing methods (see earlier) are not applicable outside designated establishments (s. 2(7))
(d) Before the release of a free-living animal it must have been certified as fit by a veterinary surgeon or other suitably qualified person (personal licence condition 9) (See Appendix 3 (Note 16))
(e) The Protection of Animals Acts do not apply to free-living non-domesticated animals but temporary restraint may well render them captive (see later, Chapters 3 and 7 and Appendix 3 (Note 8)). However, the Wildlife and Countryside Act 1981 and other wildlife legislation (see Chapter 7) must be observed in respect of protected species, methods of capture and licensing requirements

LEGISLATION INTERACTING WITH THE ANIMALS (SCIENTIFIC PROCEDURES) ACT 1986

The following Acts make specific exemptions from provisions which would otherwise render a regulated procedure an offence under those Acts. Schedule 3 of the 1986 Act provides for the amendment of these Acts so that they no longer refer to the Cruelty to Animals Act.

TABLE 1 Timetable for Implementation of the Animals (Scientific Procedures) Act 1986

Note:	Schedule 4 of the 1986 Act provides for the continuation of licences, experiments and registered premises authorised by the 1876 Act until expiry or replacement by authorisation under the new Act. Provision (summarised below) has been made for the transition to be phased over five years, the Act to be fully implemented by 1 January 1992 in accordance with Home Office (1986d)
Oct to Dec 1986	Issue of personal and project licences and of certificates to scientific procedure establishments in respect of procedures not regulated under the 1876 Act (e.g. antiserum production, passaging of tumours or parasites and the breeding of animals with harmful genetic defects) and new projects. Code of Practice on care and accommodation of laboratory animals to be produced
Nov 1986	Animal Procedures Committee, Panel of Assessors, Representation system } established
1 Jan 1987	Commencement of Animals (Scientific Procedures) Act 1986 (most parts) Licences and certificates must be held for work and premises not covered by 1876 Act New statistical system including data for Council of Europe statistics New records system for all scientific procedure establishments and for other designated breeding and supplying establishments
1 Jan to 3 March 1987	Issue of certificates of designation for all scientific procedures establishments formerly registered under 1876 Act
1 Apr 1987	All scientific procedure establishments to hold certificates of designation
Jan 1987 to Dec 1988	Project licences to be issued for 1876 authorised work: commercial concerns by 30 Sept 1987 others by 1 July 1987 to 31 Dec 1988
Jan 1987 to Dec 1991	Personal licences to be phased in, initially on submission for renewal or amendment
Jan 1988	Personal licence annual returns under new statistical system to be submitted for first year of Act Certificated designated establishments to pay fees annually
Jan 1989	All 1876 Act-authorised work to be covered by a project licence
Jan to Dec 1989	Issue of certificates of designation to breeding and supplying establishments

TABLE 1 Continued

Jan 1990	All breeding and supplying establishments to hold certificates of designation Section 10(3) in force: cats and dogs to be bred at and obtained from designated breeding establishments; other Schedule 2 animals to be obtained from designated breeding or supplying establishments New annual return submitted (for 1989) by project licence holders for first time
Jan 1990 to Dec 1991	All 1876 licences to be replaced by personal licences
Jan 1992	All authorisation by licences and certificates to be under 1986 Act

TABLE 2 Animals (Scientific Procedures) Act. A Brief Outline of its Main Provisions

	MAIN PROVISION
s. 3	"No person shall apply a *regulated procedure* to a *protected animal* unless he holds a personal licence, [the procedure is part] of work specified in a *project licence* [and] carried out [at] a *place* specified in the personal licence and the project licence."
	PROTECTED ANIMAL (PA)
s. 1	"any living vertebrate [except] man", sub-phylum Vertebrata of Phylum Chordata includes foetal, larval and embryonic forms which are protected: during the latter half of its gestation period (mammal, bird, reptile) once capable of independent feeding (other species) when living: "until permanent cessation of respiration and circulation or the destruction of its central nervous system"
	REGULATED PROCEDURE (RP)
s. 2	"any experimental or other scientific procedure applied to a protected animal which may cause that animal pain, suffering, distress or lasting harm." Includes animals bred with genetic defects. Excludes: marking animals provided that no more than momentary pain or distress is caused testing under the Medicines Act 1968 humane killing using methods permitted by Schedule 1 recognised veterinary, agricultural or animal husbandry practice
s. 3	PERSONAL LICENCE (PL) Issued to an individual to authorise: protected animals, procedures, place Must also be authorised by project licence

s. 5 PROJECT LICENCE (PrL)
Issued to individual with overall responsibility for project
Authorises project, protected animals, procedures, place and must satisfy criteria, e.g. prevention, diagnosis or treatment of disease, etc., in man, animals or plants, advancement of biological knowledge
Home Secretary must weigh adverse effects on animals against benefits of project
Cats, dogs, primates and Equidae may only be used if no other species is suitable

s. 6 DESIGNATED ESTABLISHMENTS — CERTIFICATES
SCIENTIFIC PROCEDURE ESTABLISHMENTS (SPE)
(i) CERTIFICATE held by person in position of authority
(ii) Specifies:
(NP) named person responsible for day to day care of PAs
(NVS) named veterinary surgeon (or suitably qualified person)
(iii) Health and welfare of an animal causing concern:
(a) NP must notify PL } holder or if not possible arrange
(b) NVS must notify PL } euthanasia
(c) NP may notify NVS
(d) NVS may notify HOI
(e) NP may notify HOI
(HOI: Home Office Inspector)

s. 7 BREEDING ESTABLISHMENTS (BE) } (ii) and (iii) above apply
SUPPLYING ESTABLISHMENTS (SE) } except for notification of PL holder

s. 10 LICENCES and CERTIFICATES
Subject to CONDITIONS including:
PL: holder must minimise pain, distress and discomfort in animals used
euthanasia: methods in Schedule 1 or PL must be used
PrL: Schedule 2, main laboratory species — only animals from BE/SE to be used

s. 14 RE-USE OF ANIMALS:
Permitted with HS consent only.
(HS: Home Secretary)

s. 15 Euthanasia at end of series of RPs if suffering (or likely) adverse effects
Subject to conditions in PrL for euthanasia after RP

s. 16 No RP in public or live on public television

s. 17 Neuromuscular drugs to be used only with express authority in PL or PrL; never as anaesthetic

s. 18 INSPECTORS
Advise HS on PL and PrL certificates
Inspect SPE/BE/SE
Report and advise HS on breach of licences or certificates
Power to require PL holder to kill PA which is suffering excessively

TABLE 2 Continued

ss. 19 and 20	ANIMAL PROCEDURES COMMITTEE Advises HS on the Act and his functions. Chairman and 12 members including scientists, a lawyer, animal welfare representation and non-licence holders "Must have regard both to legitimate requirements of science and industry and the protection of animals against avoidable suffering" in its deliberations Annual report to Parliament via HS
s. 21	INFORMATION HS to: publish guidance on licences, certificates or conditions issue or approve codes of practice on care and use of PAs publish annually information on usage of PA
s. 22–26	PENALTIES Fines and/or imprisonment for: contravention of sections 3, 7 and 14–17 disclosure of confidential information (s. 24) PrL holder permitting RPs not authorised by PrL or PL (s. 22) Revocation of licence or certificate for breach of conditions (not offence), but HS can revote vary, suspend a licence or certificate Prosecutions under this Act or Protection of Animals Act 1911 (1912 Scotland) require the consent of the DPP

Act applies to England, Wales, Scotland and N. Ireland.

Note: this summary is extremely brief and generalised. The reader must refer to the legislation and other literature.

TABLE 3 Offences under the Animals (Scientific Procedures) Act 1986

Section	*Offence*	*HOG*	*Penalty*
s. 22(1)	Contravention of s. 3 Performance of procedure not authorised by personal or project licence or at unauthorised place	11	1
s. 22(4)	Defence: reasonable belief, after due enquiry, of being authorised by project licence		
s. 22(2)	Procuring or permitting by a project licence holder, of a person under his control to perform a regulated procedure not authorised by his personal or the project licence	12	1
s. 10(4)	Except: permitted assistance	13 Annex B	

s. 22(3)(a)	Contravention of:			
	s. 7(1)	uncertified breeding establishment	15	2
	s. 7(2)	uncertified supplying establishment	15	2
	s. 14	unauthorised re-use	16	2
	s. 15	failure to kill an animal suffering, or likely to suffer, at the end of a regulated procedure or series thereof	17	2
	s. 16	public exhibition or live television broadcast of a regulated procedure	18	2
	s. 16	advertising thereof	18	
	s. 17(a)	unauthorised use of neuromuscular blocking agents	19	2
	s. 22(4)	defence: reasonable belief of being authorised by project licence		
	s. 17(b)	use thereof as an anaesthetic	19	2
s. 22(3)(b)	s. 18(3)	failure to comply with inspector's request to kill an animal suffering excessively	20	2
s. 22(5)	Protection of Animals Act 1911, s. 1 Protection of Animals (Scotland) Act 1912, s. 1		23	
s. 29(5)	Welfare of Animals (Northern Ireland) Act 1972, s. 13, s. 14 Causing cruelty or unnecessary suffering in respect of an animal at a designated establishment (see Note)			1
s. 23	Provision of false or misleading information to obtain a licence or certificate		21	2
s. 24	Disclosure of confidential information		14	1
s. 25(3)(a)	Intentional obstruction of a constable or inspector acting under a search warrant		22	2
s. 25(3)(b)	Refusal to supply name and address Giving false name or address		22	2

Note: In England and Wales, a prosecution brought under the Protection of Animals Act 1911, s. 1 and in Northern Ireland a prosecution brought under the Welfare of Animals (Northern Ireland) Act 1972 sections 13 and 14 in respect of an animal at a designated establishment (but not, for example, in fieldwork) or any prosecution under the Animals (Scientific Procedures) Act 1986 must have the prior consent of the appropriate Director of Public Prosecutions (s. 26(1), s. 29(5), HOG 25).

Maximum penalties

Penalty	*Conviction*	*Fine*		*Imprisonment*	*HOG*
1	Summary	Statutory maximum (£2000 in 1987)	and/or	Six months	23
	or on Indictment	Unlimited	and/or	Two years	
2	Summary	Fourth level on standard scale (£1000 in 1987)	and/or	Three months	24

Protection of Animals Act 1911; Protection of Animals (Scotland) Act 1912; Welfare of Animals (Northern Ireland) Act 1972

It is an offence to treat an animal cruelly or to cause it unnecessary suffering. In the latter case this can include the failure to provide food or water or necessary veterinary attention (see Chapter 3).

The Acts specifically provide that such offences cannot be applied to ''any act lawfully done under the Animals (Scientific Procedures) Act 1986''. Such activities are therefore protected from prosecution under the 1911, 1912 or 1972 Acts. Consequently, it may be deduced that this protection would not extend to regulated procedures carried out without, or beyond the scope of, a licence, since these would constitute offences under sections 3 and 22.

The consent of the Director of Public Prosecutions is required before a prosecution is brought under section 1 of the Protection of Animals Act 1911 or the Welfare of Animals (Northern Ireland) Act 1972 when it is to relate to an animal kept at a designated establishment (see Table 1) (s. 26(1), s. 29(5), HOG 25).

Protection of Animals (Anaesthetics) Acts 1954–1964

The Protection of Animals (Anaesthetics) Acts require the use of anaesthesia during surgical operations (see Chapters 3 and 6). The 1954 Act contains a specific exemption for ''any procedure duly authorised under the Animals (Scientific Procedures) Act 1986''.

Agriculture (Miscellaneous Provisions) Act 1968

It is an offence to cause or permit unnecessary pain or distress to livestock on agricultural land (see Chapter 3). There is an exemption from this for ''any act lawfully done under the Animals (Scientific Procedures) Act 1986'' or in accordance with a Ministry of Agriculture licence for scientific research. However, the Minister has never issued such a licence.

Veterinary Surgeons Act 1966

The Veterinary Surgeons Act section 19(1) reserves to those registered with the Royal College of Veterinary Surgeons the right to practise veterinary surgery (see Chapter 6). Subject to some exceptions, the unqualified practice of veterinary surgery is an offence.

Many of the procedures performed under the 1986 Act also fall within the definition of veterinary surgery given in the Veterinary Surgeons Act; consequently, section 19(4)(b) of the lattter Act provides that the restriction on veterinary practice shall not apply to "the carrying out of any procedure duly authorised under the Animals (Scientific Procedures) Act 1986" (also see earlier (Regulated procedures) for the exclusion of recognised veterinary procedures from the requirements of the 1986 Act).

Animal Health and Welfare Act 1984; Slaughter of Poultry Act 1967

The Animal Health and Welfare Act 1984 section 5 amends the Slaughter of Poultry Act 1967 and restricts the methods of slaughter which may be used for poultry (see Chapter 3). This requirement does not apply to "a procedure duly authorised under the Animals (Scientific Procedures) Act 1986". Poultry used for regulated procedures must be killed by a method approved in Schedule 1 to the 1986 Act or authorised in the personal licensee's licence.

Badgers Act 1973

The Badgers Act 1973 section 8(3) makes an exception to the offences of killing or injuring a badger for "doing anything which is authorised by the Animals (Scientific Procedures) Act 1986".

Dangerous Wild Animals Act 1976; Dangerous Wild Animals Act 1976 (Modification) Order 1984

The Dangerous Wild Animals Act 1976 section 5(4) requires the keeper of any non-domesticated species listed in the revised Schedule contained in the 1984 Order to hold a local authority licence (see Chapter 3). By section 5(4) no licence is required in respect of such animals while they are kept at an establishment designated under the 1986 Act. Once a dangerous wild animal is removed from a designated establishment a licence must be obtained (Cooper, 1978).

OTHER LEGISLATION RELEVANT TO ANIMALS USED IN RESEARCH

Dogs Act 1906

Stray dogs seized by the police may not be "given or sold for the purposes of vivisection" (s. 3(5)).

Breeding of Dogs Act 1973; Breeding of Dogs (Northern Ireland) Order 1983

The Breeding of Dogs Acts which require the licensing of breeding bitches do not apply to dogs bred at a designated breeding establishment for use in regulated purposes.

Animal Health Act 1981 (As Amended, Together with Orders Made Thereunder)

Provisions relating to matters such as disease control, importation, exportation and welfare in transportation (see Chapters 3 and 5) are also applicable to animals used in research.

The Zoonoses Order 1975 makes provision for the reporting and investigation of outbreaks of salmonellosis or brucellosis in birds, mammals or other four-footed animals. Article 7(5) provides that there is no duty to report the deliberate introduction of these organisms for research or experimental purposes provided that the animal is not used for human consumption and is disposed of without risk to human health.

Pests Act 1954

The offence for using an infected rabbit to spread myxomatosis among uninfected rabbits does not apply to "any procedure duly authorised under the Animals (Scientific Procedures) Act 1986".

EEC Regulation 3626/82; Endangered Species (Import and Export) Act 1976

The licensing requirements for the import and export of non-indigenous species are applicable to animals intended for research (see Chapter 7).

The directive on experimental animals (EEC 1986) requires UK law to provide that endangered (CITES Appendix I and Regulation Annex C1) species must not be used for regulated procedures unless:

(a) the animals are in conformity with the Regulation; and
(b) the research is aimed at preservation of the species used; or
(c) for a biomedical purpose, the species is the only one suitable

Wildlife and Countryside Act 1981

The restrictions upon taking animals from, or releasing them to, the wild (see Chapter 7) are applicable to free-living wildlife used for regulated procedures. If they are taken into captivity they also become subject to the cruelty legislation; however, the Protection of Animals Acts do not apply to free-living animals held under temporary restraint only (see Chapter 3).

LEGISLATION WHICH REQUIRES THE PERFORMANCE OF EXPERIMENTS ON ANIMALS

The following Acts, which require substances to be tested for efficacy or safety or both, necessarily involve the use of animals under the 1986 Act (Uvarov, 1985; Smyth, 1978):

Public Health Act 1936
Agriculture (Poisonous Substances) Act 1952
Consumer Safety Act 1978
Medicines Act 1968
Control of Pollution Act 1974
Health and Safety at Work etc. Act 1974
Biological Standards Act 1975
Food Act 1984

In addition, work is carried out in the UK, particularly in the field of toxicological testing of products and substances, to standards set by the legislation of other countries. While such laws have no effect within the jurisdiction of Great Britain and Northern Ireland, their relevant requirements are observed. The most widely used are probably the USA's Toxic Substances Control Act 1976 and Food, Drug and Cosmetic Act 1938 and their regulations for Good Laboratory Practice but many other countries have similar legislation (Home Office, 1979; Shillam, 1986).

Appendix to Chapter 4

CRUELTY TO ANIMALS ACT 1876

History

Prior to 1 January 1986 the British legislation relating to animals used for research was the Cruelty to Animals Act 1876. In all its years the substantive content of this statute was not once amended. From time to time there were unsuccessful efforts to replace the Act (summarised in Uvarov (1985)). However, in the early 1980s further moves towards new legislation and the government's proposals for reform which were published in *Scientific Procedures on Living Animals* (Home Office, 1983, 1985; Anon, 1985) led to the passing of the Animals (Scientific Procedures) Act 1986. The history of the Act and its implementation are described in Littlewood (1965) and Hampson (1978).

The Cruelty to Animals Act was repealed on 1 January 1986 although the transition to full working of the new Act will take several years. For this reason and for the benefit of those who have worked under the former legislation an account of the 1876 Act and its implementation is provided.

Implementation

The Act was put into effect by the Secretary of State for the Home Office whose powers and duties under the Act permitted the establishment of a body of administrative practice (supplementing the legal requirements) for the supervision and control of those performing painful experiments on living animals. A summary of Home Office practice is given in RDS (1974).

The Home Secretary was advised by an Inspectorate which comprised 15 medically or veterinarily qualified inspectors. Their duties included the inspection of registered premises and animals under experiment; they also advised the Home Secretary on applications for licences and special conditions thereon, on certificates and on the registration of premises, whether existing, new or undergoing alteration; in addition, they advised licensees and those responsible for registered premises. They assisted the Home Secretary to secure compliance with the law and reported to him any irregularities which came to their notice.

Advisory Committee

The Advisory Committee on Animal Experiments advised the Home Secretary upon proposed experiments of a novel or sensitive nature. It also, in later years, produced reports on important issues, namely the lethal dose 50 test (Home Office, 1979) and the framework for revision of the 1876 Act (Home Office, 1981).

Accounts of the Act and its application are to be found in, *inter alia*, RDS (1974, 1979), O'Donoghue (1980), Home Office (annual), Cooper (1981), Home Office (1981), Home Office (1983), Blackman (1985), Uvarov (1985) and Heath (1986).

For statistical data on animal experiments reference should be made to those prepared annually by the Home Secretary (Home Office, annual) and until 1977 presented to Parliament but latterly issued as a Command Paper.

The provisions of the Act and its implementation are described below and

TABLE A1 Authorisation by Licence and Certificate under the Cruelty to Animals Act 1876

Authorisation			*Anaesthesia*	*Recovery*	*Disposal*
Licence only			Yes	No	
Licence +	Certificate F	Horses	Yes	No	
Licence +	Certificate A		No	Yes	Animal returned to stock if fully recovered
Licence +	Certificates A + E	Cats and dogs	No	Yes	Animal returned to stock if fully recovered
Licence +	Certificates A + F	Horses	No	Yes	Animal returned to stock if fully recovered
Licence +	Certificate B		Yes	Yes	Animal killed at the end of the experiment
Licence +	Certificates B + EE	Cats and dogs	Yes	Yes	Animal killed at the end of the experiment
Licence +	Certificates B + F	Horses	Yes	Yes	Animal killed at the end of the experiment
Licence +	Certificate C	Lectures	Yes	No	
Licence +	Certificates C + F	Horses	Yes	No	
Licence +	Certificate D	Obsolete			

Written authority was required for the use of non-human primates.

TABLE A2 The Cruelty to Animals Act 1876: A Summary of its Requirements and Home Office Implementation

	Section	*Certificate*	*Condition*	*Requirement*
s. 1	Title			
s. 2	Experiments on animals calculated to give pain may only be performed under restrictions imposed by Act			
	Breach of s. 2 = offence	No successful prosecutions; see s. 8 revocation		
s. 3	Restrictions on performance of experiments (3P's):			
	(1) Promoting physiological knowledge Preventing suffering Prolonging life			
	(2) Under licence (3) Under anaesthesia (4) Non-recovery	Every experiment requires licence		
	(5) No demonstrations			
	(6) Not for manual skill			
	Provisos:			
	(1) For demonstration in lectures	Certificate C	6.	Animal to be killed, at end of experiment by or in presence of licensee
	(2) Without anaesthesia	Certificate A	3. 4.	Pain condition No procedure more severe than simple inoculation
	(3) Under anaesthesia with recovery	Certificate B	3. 5.	Pain condition Anaesthesia and antiseptic precautions + failure → killed immediately

	(4) Re-testing	Certificate D, obsolete		
s. 4	Curari not an anaesthetic		7.	No muscle-relaxant drugs to be used without HO permission, 48 h notice to inspector
s. 5	Cats and dogs Horses Certificate given only if no other species suitable	Certificates E and EE Certificate F (L + F; A + E or F; B + EE or F)		
s. 6	Public exhibitions of experiments illegal	cf. Certificate C; bona fide colleagues		
s. 7	Registration of premises: experiments to be performed in registered premises; Certificate C experiments only on registered premises		1.	States place of "availability"
s. 8	HO may license any person for any duration and revoke licence as HO thinks fit May add conditions to licence for putting Act into effect and consistent with it	Revocation Reprimand Warning	10. Special conditions	Primates e.g. supervision
s. 9	HO may require reports of experiments		8. 9.	Written record of experiments open to inspection; Annual Return to HO Publications involving experiments

TABLE A2 Continued

	Section	Certificate	Condition	Requirement
s. 10	Registered premises to be inspected HO power to appoint inspectors		14.	Inspectors' duties: inspect experiments and animals, registered premises, advise licensees, read papers, review films
s. 11	Statutory signatories Presidents of learned medical societies + professors of scientific/medical subjects: sign licence application grant certificate Licensee sends copy of certificate to HO Certificate not available for 1 week + HO power to disallow or suspend certificate		2.	No experiments before notification from HS that certificate is not disallowed
s. 12	High Court Judge may grant licence or certificate if experiment is essential for justice in a criminal case	Probably only in one case — 1980		
ss. 13–20	Legal proceedings			
s. 20	Prosecution against a licensee only with consent of HS			
s. 21	Act does not apply to invertebrates			

the licensing requirements are set out in Table A1. Table A2 shows the correlation between these and the standard licence conditions.

Definitions

The key provision of the Act (and the chief of the three offences created by the Act) was section 2, which provided that:

> A person shall not perform on a living animal any experiment calculated to give pain, except subject to the restrictions imposed by this Act

The Act itself offered little help in interpreting the various elements of this statement and there were no judicial decisions to provide authoritative definitions. However, working applications evolved and can be examined in detail as follows.

Living animal

The Act applied to all animals except invertebrates (s. 22), including such of their young as are capable of an independent existence; the word living was applied to an animal while "it is breathing and its heart is beating and its cerebrum and basal ganglia are intact" (RDS, 1979).

Experiment

An experiment was taken to mean a procedure, the outcome of which is not known in advance. It includes "any procedure which may interfere with the normal well-being of a vertebrate animal other than killing" (RDS, 1974). This was very broadly interpreted to cover not only research on the boundaries of existing knowledge but also the quality control testing of drugs and other medicinal products for safety and efficacy in the course of routine production. By comparison, the production of substances derived from animals (e.g. blood products such as therapeutic antisera) and the passaging of tumours were not considered subject to the Act. Routine procedures such as the induction of anaesthesia or euthanasia, which often take place in the course of an experiment, were not classified as experiments.

The Act did not apply to stock animals (i.e. animals destined for use in an experiment) or control animals not undergoing potentially painful procedures; however, positive controls (e.g. infected animals) or ones in which intrusive techniques were used to measure the normal for comparison were regarded as experimental animals. Consideration of the totality of an experiment brought within the scope of the 1876 Act some procedures which in earlier years would have been considered outside it. For example, the surgical preparation of an animal by hypophysectomy or thyroidectomy did not, of itself, advance physiological knowledge and hence would be outside the Act; but such operations were considered part of the experiment if they were required to permit subsequent observations such as the evaluation of a new substance, the observation of control animals to validate the study or the use of new methods of research. Likewise, removal of tissues or organs from live animals, although the material and not the animals was the subject of investigative work (as in *in vitro* experiments), also required appropriate authority. If the animal was killed first and organs were removed no licence was needed but if, for reasons of obtaining sufficiently fresh material, the organ was removed from a live animal which had been anaesthetised or was perfused with fixative a licence was required.

Calculated to cause pain
The phrase included any reasonable likelihood that the procedure might materially interfere with an animal's health or well-being. Thus, not only extensive surgical procedures but also minor interferences were included. Pain had always to be evaluated without taking into consideration the use of anaesthesia.

Justification of Experiments

The Act (s. 3) required that "The experiment must be performed with a view to the advancement by new discovery of physiological knowledge or of knowledge which will be useful for saving or prolonging life or alleviating suffering". In short, experiments had to satisfy at least one of what might, for simplification, be called "the three Ps", namely, physiological knowledge, preservation of life, or the prevention of suffering. This might be for the benefit of either humans or animals.

Applications for a licence or certificates authorising work carried out within the Act had to include a statement indicating the way in which one or more of "the three Ps" would be achieved.

Licence and Certificates

Any person performing an experiment under the Act had to hold a Home Office licence. Section 3(3) of the Act demanded that, throughout an experiment, an animal should be sufficiently anaesthetised to prevent its feeling pain. If such pain was likely to continue after the anaesthesia had ceased or if the animal had undergone some serious injury the animal had to be killed before it recovered from the anaesthetic. In effect, the *prima facie* requirements of anaesthesia and non-recovery applied to acute experiments carried out under licence only (see Table A1).

In the many situations in which the requirements for anaesthesia and non-recovery would defeat the purpose of an experiment, certificates were issued to dispense with one or both of these restrictions.

Certificate A (see Table A1) authorised a licence holder to perform without anaesthesia minor procedures no more severe than, for example, a simple inoculation. At the end of the experiment, the animal could be returned to stock provided that it was unharmed.

Certificate B (see Table A1) was applied to experimental work carried out under anaesthesia and permitted recovery of the animal from the anaesthetic until such time as the purpose of the experiment had been achieved whereupon it had to be killed humanely.

Work carried out under Certificates A and B was subject to the pain condition (i.e. condition 3 of the licence; see below).

Other certificates were required to authorise the use of cats, dogs and horses and to perform an experiment in the course of teaching (see Table A1). Written permission from the Home Office was required for the use of primates or neuromuscular blocking agents.

Experiments were not allowed to be performed purely as a means to obtain manual skill (s. 3(6)). A distinction had to be drawn between developing a new technique and learning an existing one. Under the 1876 Act those who wished to acquire established surgical skills had to use human patients or dead animals or study abroad. Under the 1986 Act this restriction has been relaxed in respect of training in microsurgery.

When a licensee proposed to commence a new line of research or a different procedure it was his duty to ensure that the terms of his existing licence and certificates covered the proposed experiments.

Licence Conditions

A Home Office licence was issued subject to conditions. Standard conditions were printed in every licence and further special conditions were inserted in individual licences. Table A2 correlates the standard conditions with the provisions of the 1876 Act.

The standard conditions served to implement the requirements and spirit of the Act (see Table A2). The chief ones were as follows:

Condition 1 A licence holder might only perform experiments at the registered premises (see below) named in that licence

Condition 2 Disallowance (see below)

Condition 3 If during a procedure carried out under licence and Certificates A and B an animal suffered pain which was either severe *or* likely to endure, the animal had to be painlessly killed once the main result of the experiment had been achieved: if, however, pain occurred which was both severe *and* likely to endure, the animal had to be killed immediately.

It was the responsibility of the licensee to make that assessment and act accordingly. In addition, a Home Office inspector (see below) could require the euthanasia of any experimental animal which appeared to be suffering considerable pain.

Condition 4 Restricted Certificate A to very minor operative procedures

Condition 5 Required the use in Certificate B experiments of adequate anaesthesia and antiseptic precautions; should these fail and pain result the animal had to be killed under anaesthesia

Condition 6 Provided that animals used under Certificate C had to be killed at the end of an experiment by or in the presence of the licensee

Condition 7 Stipulated that special permission was required for the use of curari and other neuromuscular blocking agents

Condition 8 Required records of experimental work to be kept

Condition 9 Required that copies of publications describing licensed experiments had to be supplied to the Home Office

Condition 10 Required that written permission of the Home Office had to be obtained to perform experiments on primates

Special conditions could be imposed in individual licences and these most frequently restricted work to specified procedures or required the supervision of inexperienced and non-graduate licensees or imposed veterinary supervision in the case of novel techniques. A worker from overseas had to have a supervisor who gave a written undertaking to the Home Office that he would ensure that the licensee understood, and complied with, the Act.

When proposed experiments presented particular problems (e.g. procedures which were new or were considered sensitive to public opinion), caused special risks to the animals used or involved endangered species, it was necessary to provide the Home Office with a detailed programme of the proposed work.

Application and Grant of Licences and Certificates

Applications for licences and certificates were made on special forms. A general description of the procedures to be authorised by Certificate A and a detailed description of a specific experiment for Certificate B was required. In addition the applicant had to justify the proposed work in terms of the "three Ps" mentioned earlier and give reasons, in accordance with section 5, for any intended use of cats, dogs and equids rather than other species. Permission also had to be obtained if primates were involved.

The licence application had to be signed by two "statutory signatories", who were persons of a standing stipulated by the Act, namely the Presidents of the Royal Society, of certain Royal Colleges or of the General Medical Council, together with a professor of a medical subject or appropriate biological science.

The application was then forwarded to the Home Office which, if satisfied, granted and issued the licence.

In the case of a certificate, however, the application form also comprised the certificate itself which was granted by the statutory signatories and not by the Home Office.

In view of this procedure, the Act (s. 11) provided that a copy of the certificate had to be sent to the Home Secretary, that it was not to come into force until seven days after such submission and that the Home Secretary might disallow or suspend it. Condition 2 of a licence provided that a licensee could not use a certificate before receiving notification from the Home Office that it had not been disallowed. In addition, of course, a licensee had to be in possession of his certificates and licence (not awaiting their issue or renewal) before commencing experimental work. Licences and certificates had to be returned to the Home Office for renewal (usually after five years), to obtain additional, or a change of, availability (see later: Registered premises), when experimental work ceased or when additional authority in the form of new certificates or requests for change or cancellation of any of the special conditions of the licence were sought.

Records

Condition 8 of a licence required a licensee to keep records. These comprised the following:

(a) Day-to-day records of each experiment performed; this record had to contain the dates of experiments and sufficient information to enable the licensee to compile (c)

(b) Cage cards which identified the licensee, the experiment and the date it was commenced and the animals used; a card had to be identifiable with the relevant day-to-day records
(c) An annual report to the Home Office of experiments performed in the year to which it related. Licensees' end-of-year reports were, as they still are under the 1986 Act, used by the Home Office to compile the Home Secretary's annual return to Parliament of *Statistics of Experiments on Living Animals* (Home Office, annual)

Registered Premises

The Home Office required that experiments under the 1876 Act were, with limited exceptions, performed at registered premises. Further, a licence holder was restricted to working at the registered premises named in that licence; it was said that the licensee had "availability" at such premises. Before commencing work at additional premises the appropriate availability had to be added to the licence.

Special permission was granted occasionally for work which could only be done outside registered premises, e.g. in the field, subject to 72 hours' notice being given of the place and time at which such permission was to be exercised.

Experimental animals could not be moved out of registered premises or transferred to other premises, registered or otherwise, without Home Office consent. A written application had to be submitted giving details of the premises involved, the means of transport and the provision for compliance with other legislation relating to, for example, welfare in transport or disease control (see Chapters 3 and 5).

Registered premises were subject to inspection by Home Office inspectors. Construction of, or alterations to, registered premises had to be approved by the Home Office. RDS (1974) provided guidance on caging, environmental and other standards. Other publications also supplied guidelines for animal house management (e.g. Biological Council (1984), COE (1986b)) which will continue to be relevant to the new legislation. Individual institutions had (and will continue to have, after the change in legislation) in-house rules for good management and welfare of animals.

Responsibility of the Licensee

A licensee was personally responsible for the work and the animals used under the 1876 Act. A licence and certificates were issued to an individual authorising only such work as fell within their scope.

The delegation of the authority of a licence to another person was forbidden although this did not preclude the collaboration of properly authorised licensees in a conjoint experiment or the provision of mechanical duties, such as restraint of an animal or administration of a medicated diet, by an unlicensed person. Procedures outside the Act, such as routine anaesthesia or euthanasia, could be performed by persons other than the licensee, albeit without protection of the exemption from the Protection of Animals Acts and the Veterinary Surgeons Act (see earlier in this chapter).

The welfare of animals used in an experiment was the personal responsibility of the licensee using them. This included the duty to ensure that the licence conditions, especially those relating to pain, were fulfilled. A licensee who was not able to give personal attention to an animal nevertheless had to be available or have left instructions to authorise implementation of these conditions (RDS, 1974). An animal house had to have standing orders which authorised treatment or euthanasia if the licensee was not available.

Some institutions also had (and continue to have under the 1986 Act) user, research review or ethical committees which examined the merits of proposed animal usage — although there was no statutory obligation for this (Britt, 1983, 1985a, 1985b; Cooper, 1985; LAEG, 1986).

Demonstration of Animal Experiments

The Cruelty to Animals Act 1876 section 6 prohibited the demonstration to the general public of experiments covered by the Act. Nevertheless, the first proviso to section 3 permitted the performance of animal experiments in the course of lectures to learned scientific societies or when teaching students. The procedure had to be authorised by licence together with Certificate C and the animal had to be anaesthetised throughout the experiment and humanely killed by, or in the presence of, the licensee at the end of the experiment.

The Act did not forbid the observation of ongoing research by those with a *bona fide* scientific interest in it.

Enforcement

The Home Office inspectors were required to report any failure to comply with the Act or licence conditions (for their general duties, see earlier). The Act provided for prosecution for three offences: the performance of an experiment otherwise than in accordance with the Act (s. 2), the demonstration of an experiment to the general public (s. 6) and the obstruction of the police in certain duties (s. 13). No prosecution against a licensee could be instigated without the consent of the Home Secretary.

Few prosecutions were commenced or considered in the lifetime of the Act, none successfully. While some breaches had become time-barred by the time they came to the notice of the authorities, others were referred to the Director of Public Prosecutions to advise on the need for prosecution. Over 60% of the infringements in 1983 were referred although no prosecutions were recommended. In 1982 the police were instructed by the Director of Public Prosecutions to issue a caution in one case (Home Office, annual).

The Home Secretary had the power (provided in section 8 of the Act) to revoke a licence for a breach of the Act or the conditions of a licence (the latter was not an offence). In addition, a licensee could be warned, admonished or called to the Home Office to explain a breach; revocation of a licence was rare but not unknown (Home Office, annual). Most infringements involved the performance of experiments before a licence or certificate was granted, after it had expired or an excess of its authority; other examples were unlicensed assistance with experimental work or performing animal experiments at unregistered premises or at registered premises without availability.

REFERENCES

Anon (1980). *Report of the Select Committee of the House of Lords on the Laboratory Animals Protection Bill*. Vol. 1. (Report, Minutes, Proceedings) Cmnd 246-I; Vol. 2 (Minutes & Evidence & Appendices) Cmnd 246-II. HMSO, London.

Anon (1985). Experimental animals: tighter controls on procedures proposed. *Veterinary Record* **116**, 554–555.

Anon (1986). Animals (Scientific Procedures) Act 1986: 1. The Animal Procedures Committee. *Frame News* **11**, 6–7.

Anon (1987). Legislation and laboratory animals. In *The UFAW Handbook on the Care and Management of Laboratory Animals* (Poole, T., ed.). Longman, London.

Balls, M. (1986). Animals (Scientific Procedures) Act 1986: the Animal Procedures Committee. *Alternatives to Laboratory Animals* **14**, 6–13.

Biological Council (1984). *Guidelines on the Use of Living Animals in Scientific Investigations*. Biological Council, London.

Blackman, D.E. (1985). Legal and ethical constraints on animal experimentation. In *Animal Experimentation: Improvements and Alternatives* (Marsh, N. and Haywood, S., eds). Fund for Replacement of Animals in Medical Experiments, Nottingham.

Britt, D.P. (1983). The potential role of local ethical committees on the moderation of experiments on animals in Britain. *International Journal for the Study of Animal Problems* **4**, 290–294.

Britt, D.P. (1985a). Animal research review committees — a means of improving animal experiments? In *Animal Experimentation: Improvements and Alterna-*

tives (Marsh, N. and Haywood, S., eds). Fund for the Replacement of Animals in Medical Experiments, Nottingham.

Britt, D.P. (1985b). *Research Review (Ethical) Committees for Animal Experimentation*. UFAW, Potters Bar.

BVA/RCVS (1986). *Notes for establishments employing veterinary surgeons as advisors under the Animals (Scientific Procedures) Act 1986*. British Veterinary Association and Royal College of Veterinary Surgeons, London.

BVA/RCVS (1987). Guidelines for veterinary surgeons employed in scientific procedure establishments and breeding and supplying establishments. *Veterinary Record* **120**, 17–19.

CIOMS (1985). *International Guiding Principles for Biomedical Research Involving Animals*. Council for International Organizations of Medical Sciences, Geneva.

COE (1986a). *European Convention for the Protection of Vertebrate Animals used for Experimental and other Scientific Purposes*. Council of Europe, Strasbourg.

COE (1986b). Guidelines on Accommodation and Care of Animals. Appendix A to the *European Convention for the Protection of Vertebrate Animals used for Experimental and other Scientific Purposes*. Council of Europe, Strasbourg.

COE (1986c). Statistical tables and explanatory notes for their completion in fulfilment of the requirements in Articles 27 and 28 of the Convention. Annex B to the *European Convention for the Protection of Vertebrate Animals used for Experimental and other Scientific Purposes*. Council of Europe, Strasbourg.

Cooper, J.E. (1985). Ethics and laboratory animals. *Veterinary Record* **116**, 549–595.

Cooper, J.E. (1987). Euthanasia of captive reptiles and amphibians: Report of UFAW/WSPA working party. In *Euthanasia of Animals*. Universities Federation for Animal Welfare, Potters Bar.

Cooper, J.E., Ewbank, R., Platt, C. and Warwick, C. (1986). Euthanasia of reptiles and amphibians. *Veterinary Record*, **119**, 484.

Cooper, M.E. (1978). The Dangerous Wild Animals Act 1976. *Veterinary Record* **102**, 475–477.

Cooper, M.E. (1981). The law relating to animal experiments. In *The Law for Biologists*. Institute of Biology, London.

EEC (1986). Council Directive of 24 November 1986 on the approximation of laws, regulations and administrative provisions of the Member States regarding the protection of animals used for experimental and other scientific purposes. Official Journal Vol. 29, No. L358 of 18 December 1986. Council of Ministers of the European Communities, Brussels.

Fox, H. (1986). Guidelines for the Use of Animals in Research. *Animal Behaviour* **34**, 315–318.

Griffin, G. (ed.) (1985). Supplementary White Paper proposals. Summary of government proposals . . . (as in Cmnd 8883 and Cmnd 9521). *Frame News* **6**, 6–7.

Hampson, J.E. (1978). Animal Experimentation 1876–1976: Historical and Contemporary Perspectives. PhD Thesis, University of Leicester, Leicester.

Heath, M. (1986). British law relating to experimental animals — its provisions and restrictions. *Animal Technology* **37**(2), 131–135.

Home Office (annual). *Statistics of Experiments on Living Animals in Great Britain*. HMSO, London.

Home Office (1979). *Report on the LD50 Test*. Home Office, London.

Home Office (1981). *Report on the Framework of Legislation to Replace the Cruelty to Animals Act 1876*. Advisory Committee on Animal Experiments, Home Office, London.

Home Office (1983). *Scientific Procedures on Living Animals.* Cmnd 8883. HMSO, London.

Home Office (1984). *Statistics of Experiments on Living Animals in Great Britain 1983.* HMSO, London.

Home Office (1985). *Scientific Procedures on Living Animals.* Supplementary White Paper, Cmnd 9521. HMSO, London.

Home Office (1986a). *Home Office Guidance on the Operation of the Animals (Scientific Procedures) Act 1986.* Home Office, London.

Home Office (1986b). *Guidance on Record Keeping by Designated Scientific Procedure Establishments, Project Licence and Personal Licence Holders.* Home Office, London.

Home Office (1986c). *Animals (Scientific Procedures) Act 1986. Return of Procedures 1987. Explanatory Notes and Code Lists.* Home Office, London.

Home Office (1986d). *Guide to the Transitional Arrangements for the Implementation of the Animals (Scientific Procedures) Act 1986.* Home Office, London.

Hovell, G.J.R. (1985). Council of Europe animal protection legislation — the role of ICLAS. In *The Contribution of Laboratory Animal Science to the Welfare of Man and Animals: Past, Present and Future* (Archibald, J., Ditchfield, J. and Rowsell, H.C., eds). Gustav Fischer, Stuttgart and New York.

IAT (1987). Draft Guidance Notes relating to Persons Specified in Certificates of Designation to be Responsible for the Day-to-Day Care of Protected Animals kept at Designated Establishments under the Animals (Scientific Procedures) Act 1986. Institute of Animal Technology, Oxford.

LAEG (1986). *A Colloquium: Implementing Ethical Review Committees for Animal Experiments in Britain.* Animal Ethical Committee, Liverpool.

Littlewood, Sir S. (1965). *Report of the Departmental Committee on Experiments on Animals.* Cmnd 2641. HMSO, London.

Morton, D.B. and Griffiths, P.H.M. (1985). Guidelines on the recognition of pain, distress and discomfort in experimental animals and an hypothesis for assessment. *Veterinary Record* **116**, 431–436.

O'Donoghue, P.N. (1980). Animal experiments, personal responsibility, and the law. In *Handbook for the Animal Licence Holder* (Wyatt, H.V., ed.). Institute of Biology, London.

RS/UFAW (1987). *Guidelines for the Care and Use of Animals Required for Scientific Purposes.* Royal Society, London, and Universities Federation for Animal Welfare, Potters Bar.

RDS (1974). *Guidance Notes on the Law Relating to Experiments on Animals.* Research Defence Society, London.

RDS (1979). *Notes on the Law Relating to Experiments on Animals in Great Britain.* 4th edn. Research Defence Society, London.

Richards, M.A. (1986). The Animals (Scientific Procedures) Act 1986. *The Society For General Microbiology Quarterly* **13**(4), 102–104.

Sanford, J., Ewbank, R., Molony, V., Tavernor, W.D. and Uvarov, D. (1986) Guidelines for the Recognition and assessment of pain in animals. *Veterinary Record*, **118**, 334–338.

Shillam, K.W.G. (1986). Good Laboratory Practice Regulations: the first ten years. *ICLAS Bulletin* **59**, 20–24.

Smyth, D.H. (1978). *Alternatives to Animal Experiments.* Scolar Press and Research Defence Society, London.

Tuffrey, A.A. (1987). *Laboratory Animals: An Introduction for New Experimenters.* Wiley, Chichester.

Uvarov, O. (1985). Research with animals: requirement, responsibility, welfare. *Laboratory Animals* **19**, 51–75.

Zimmermann, M. (1984). Ethical considerations in relation to pain in animal experimentation. In *Biomedical Research Involving Animals* (Bankowski, Z., and Howard-Jones, N., eds). Council for International Organizations of Medical Sciences, Geneva.

RECOMMENDED READING

See Chapter 1 for literature generally applicable.

Anon (1876). *Report of the Royal Commission on the Practice of Subjecting Live Animals to Experiments.* C. 1397. Parliamentary Papers, London.

Anon (1912). *Final Report of the Royal Commission on Vivisection.* Cd. 6114. Parliamentary Papers, London.

Anon (1984). Report of the working party on courses for animal licensees. *Laboratory Animals* **18**, 209–220.

Brook, M. (1976). Animal experiments: a personal view. *Biologist* **23**, 8–11.

IOB (1982). Animal experimentation. *Biologist* **29**(1), 43–44.

IOB (1986). Animal Acts. *Biologist* **33**(5), 245–246.

Lane-Petter, W., Fell, H.B. and Mellanby, K. (1977). Animal Experiments. *Biologist* **24**, 229–235.

Orlans, F.B. (1986). Classification system for degree of animal harm. *Scandinavian Journal of Laboratory Animal Science* **13**, 93–97.

Porter, A.R.W. (1975). The law and animal experimentation. In *An Introduction to Experimental Surgery*. (Boer, J. De., Archibald, J. and Downie, H.G., eds). Excerpta Medica, Amsterdam.

Rankin, J.D. (1982). The present system for controlling experiments on animals in Britain. In *Experiments on Living Animals*. British Association for the Advancement of Science, London.

Remfry, J. (1977). The Control of Animal Experimentation. In *Das Tier im Experiment*. (Weihe, W.H., ed.), Hans Huber, Bern.

Short, D.J. and Woodnott, D.P. (1969). The law and laboratory animals. In *The I.A.T. Manual of Laboratory Animal Practice and Techniques*, 2nd edn. Crosby and Lockwood, London.

UFAW (1974). *Animals and the Law.* Universities Federation for Animal Welfare, Potters Bar.

UFAW (1977). *The Welfare of Laboratory Animals, Legal Scientific and Humane Requirements*. Universities Federation for Animal Welfare, Potters Bar.

Vine, R.S. (1968). Requirements of the Home Office. *Laboratory Animals Symposia* **1**, 23–27.

Vine, R.S. (1976). Cruelty to Animals. *New Scientist* **71**, 588.

Warren, A.G. (1970). The law and laboratory animals. In *Nutrition and Disease in Experimental Animals* (Tavernor, W.D., ed.). Baillière Tindall & Cassell, London.

Williams, P.C. (1965). The Littlewood Report or the civil servants' charter. *Institute of Biology Journal* **12**(4), 141–145.

5 Animal Health

Then, in July 1714, a few weeks before George I came to the throne, a report reached those in authority that cattle were dying at Islington from a mysterious condition suggestive of plague . . . The Commissioners consulted with several cow-doctors (cow leeches) and . . . decided the malady was, in truth, the much dreaded cattle-plague . . . recommendations . . . consisted of immediately destroying and burning all infected cattle washing and disinfecting all cow byres, by burning pitch, tar and wormwood and leaving them empty for three months . . .

(''Animal Health. A Centenary 1865–1965'', Ministry of Agriculture, Fisheries and Food. © British Crown Copyright 1965)

GENERAL

Animal health on farms and elsewhere is of the utmost importance to the national economy and public welfare. The main emphasis of the relevant legislation is upon agricultural animals but in certain circumstances it affects other species and applies to situations not only on the farm and in the home but in zoos, research laboratories and field studies upon free-living animals.

The Animal Health Act 1981 as amended by the Animal Health and Welfare Act 1984 empowers MAFF to make orders thereunder to control disease in animals. This involves measures to limit outbreaks, to promote the eradication of disease by supervising farms, markets and the movement of animals and to restrict the importation of animals. The legislation is administered by the MAFF veterinary service (Animal Health Division) and enforced by local authorities.

The animal health legislation deals with diseases designated as ''notifiable'' such as anthrax and foot-and-mouth, the eradication of diseases such as tuberculosis and brucellosis and the control of zoonoses, the provision of veterinary services, the control of movement of animals, the cleansing and disinfection of premises and vehicles, the slaughter of, and compensation for, animals infected with a notifiable disease, and the destruction of wildlife in the course of disease control. Special powers are granted for the control of rabies. There is also provision for the restriction of importation of animals and for their welfare during slaughter and transportation; these matters are dealt with in this chapter and in Chapter 3 respectively.

The legislation is enforced by local authorities and administered by MAFF which provides a nationwide network of administration and advisory and veterinary field and laboratory services.

CONTROL OF SPECIFIC DISEASES

Notifiable Diseases

Orders made under the Animal Health Act 1981 specify many diseases (some of which are not currently present in Britain) as 'notifiable'. Legislation and the species to which it applies are set out in Table 1.

These orders affect species other than those normally found on the farm. For example foot-and-mouth disease is known to occur in deer and the relevant order specifically mentions elephants.

Likewise, the orders apply to animals even if they are not kept for agriculture. Thus, movement permits will be required to transfer an animal of a species subject to an order even if it is being transferred between zoos or research institutions.

Although these orders vary in detail they normally provide for the following:

(a) Notification of a suspected case of the disease to the police or local veterinary division of MAFF
(b) Isolation of affected animals
(c) Declaration of infected areas
(d) The control of movement of people, animals and vehicles within the area
(e) The cleansing and disinfection of affected premises and vehicles
(f) The slaughter or treatment of affected or in-contact animals
(g) Compensation for compulsory slaughter (Aujeszky's disease, tuberculosis and brucellosis only)

Special provisions are made in respect of rabies, including power to order the vaccination, confinement and control of mammals and the destruction of wildlife living within an infected area.

There is also power in the Animal Health Act to make orders permitting the destruction of wildlife to prevent the spread of other diseases. Such provisions have in the past been used to authorise the killing of badgers thought to transmit tuberculosis to cattle in certain parts of the west of England.

TABLE 1 Notifiable Diseases

Order and relevant disease	*Species to which order applies*
African Swine Fever Order 1980	Swine
Anthrax Order 1938	Horses, asses, mules
	Any four-footed mammal kept in captivity (except mammals kept in a licensed pathological institute)
Aujeszky's Disease of Swine Order 1983	Cattle, sheep, goats, swine, deer, horses, dogs, cats
Bee Diseases Control Order 1982	Bees
American foul brood	
European foul brood	
Varroasis	
Brucellosis (England and Wales) Order 1981	Bovine animals: bulls, cows, heifers, calves, but not steers
Brucellosis (Scotland) Order 1979	
Abortion or premature calving	
Brucellosis Melitensis Order 1940	Cattle, sheep, goats, swine, horses, asses, mules
Cattle Plague Order 1928	Cattle, sheep, goats
	All other ruminating animals
	Bovine animals
Diseases of Fish (Definition of "Infected") Order 1984	Fish of any kind
Diseases listed in chapter	
Dourine Order 1975	Horses, asses, mules, zebra
Enzootic Bovine Leukosis Order 1980	Bovine animals
Epizootic Lymphangitis Order 1938	Horses, asses, mules
Foot-and-Mouth Disease Order 1983	Cattle, sheep, goats, swine
	All other ruminating animals
	Elephants
	(Parts of Order affect other species)
Fowl Pest Order 1936***	Domestic fowls, turkeys, geese, ducks
Any forms of fowl pest, including	Guinea-fowl, pigeons
Newcastle disease and fowl plague	Pheasants, partridges, quails
Paramyxovirus of pigeons	
Glanders or Farcy Order 1938	Horses, asses, mules, jennets
Infectious Diseases of Horses Order 1975	Horses, asses, mules, zebra
Equine infectious anaemia	
Equine encephalomyelitis	
African horse sickness	
Pleuro-Pneumonia Order 1928	Cattle including bulls, cows, oxen, heifers, calves
Contagious bovine pleuro-pneumonia	
Psittacosis and Ornithosis Order 1953	Domestic fowls, turkeys, geese, ducks
*Chlamydiosis	Guinea-fowl, pigeons, pheasants, partridges, quails
	Doves, peafowl, swans

	Psittaciformes species including parrots parrakeets, budgerigars, love birds, macaws, cockatoos, cockatiels, conures, caiques, lories, lorikeets
Rabies (Control) Order 1974	Most mammals except man; see Schedule I of Order
Sheep Pox Order 1938	Sheep, lambs
Sheep Scab Order 1977	Sheep, lambs
Swine Fever Order 1936	Pigs
Swine Vesicular Disease Order 1972	As for Foot-and-Mouth Disease Order
Teschen Disease Order 1974	Swine
Tuberculosis (England and Wales) Order 1984 Tuberculosis (Scotland) Order 1979 Bovine tuberculosis	Bovine animals, i.e. bulls, cows, steers, heifers, calves
Warble Fly (England and Wales) Order 1982 Warble Fly (Scotland) Order 1982	Cattle of 12 weeks and older
Zoonoses Order 1975	All non-human mammals
**Salmonella organisms	Any non-mammalian four-footed beast
**Brucella organisms	Art 7: duty to report restricted to cattle, sheep, goats, pigs, rabbits, fowls, turkeys, geese, ducks, guinea-fowl, pheasants, partridges, quails (exception for research)

Only the main orders are given. In many cases there are supplementary or amending orders; for details of these and for subsequent changes see *Halsbury's Statutory Instruments* and other literature mentioned in Chapter 1.
*Scheduled and **reportable, not notifiable, diseases.
***Replaced by Infectious Diseases of Poultry Order 1986.

REPORTABLE DISEASES

The Zoonoses Order 1975 requires that outbreaks of salmonellosis and brucellosis in animals produced for human consumption must be reported to MAFF which has powers to investigate and deal with the outbreak. Occurrences in other species need not be reported but MAFF may investigate an outbreak which comes to its notice, e.g. as a result of diagnosis at a MAFF veterinary investigation laboratory. There is no requirement to report the introduction of the disease into an animal for research or experimental purposes. However, such animals must not be used for human consumption and must be disposed of without risk to human health (Lowes, 1975; Bennett, 1976; Bell, 1981).

MOVEMENT

The movement of bovine animals, sheep, goats and pigs is subject to the Movement and Sale of Pigs Order 1975 and the Movement of Animals (Records) Order 1960, as amended. Records must be kept of animals which are moved between different premises including markets.

DISEASE CONTROL IN FISH

The Diseases of Fish Acts 1937 and 1983 give MAFF powers, comparable to those under the Animal Health Acts, to deal with outbreaks of notifiable diseases in fish.

The Diseases of Fish (Definition of "Infected") Order 1984 designates the following as notifiable diseases:

Bacterial kidney disease
Infectious pancreatic necrosis (IPN)
Viral haemorrhagic septicaemia (VHS or Egtved disease)
Myxosoma cerebralis (whirling disease)
Infectious haematopoietic necrosis (IHN)
Ulcerative dermal necrosis (UDN)
Spring viraemia of carp (SVC)
Furunculosis of salmon

Notification must be made to the Fisheries Department of MAFF if it is suspected that fish at fish farms or elsewhere are infected with these diseases. MAFF has power to designate areas where there is an outbreak and to regulate the movement into and out of the area of live fish, fish eggs and fish food.

There is also provision for the registration of inland and marine fish farms and shellfish farms which will require them to maintain records.

Disease Control in Insects

The Bees Act 1980 empowers MAFF to make provision for the control of disease in bees. The Bee Diseases Control Order 1982 makes notifiable American foul brood, European foul brood and varroasis diseases and permits MAFF officers to examine bees, combs and beekeeping equipment and to test combs for these diseases. They may forbid movement and order the destruction of infected bees, combs and equipment. Alternatively they can order bees infected with European foul brood to be treated with antibiotics. In any other circumstances antibiotic or other treatment which might mask the presence of foul brood diseases is forbidden.

Insects which are damaging to plant crops, such as the Colorado beetle, are controlled by MAFF by orders made under the Plant Health Act 1967, namely the Tree Pests (Great Britain) Order 1980, the Plant Pests (Great Britain) Order 1980 and the Colorado Beetle Order 1933.

IMPORTATION

A major factor in the prevention and control of animal disease is the regulation of the movement of animals which may carry infection from one country to another.

In most countries there is a substantial network of import restrictions so that very few species of animal, their products or parts can be imported without passing through government controls. These usually include a licence to import, health certification from, and quarantine in, the country of departure, customs procedures on arrival and subsequent quarantine in the country of destination.

The orders concerned with importation of live animals into Britain are listed below, together with the species which they affect:

Importation of Animals Order 1977	Cattle, sheep, goats and all other ruminating animals (including llamas, guanacos, alpacas, vicunas, Bactrian camels, Arabian camels) and swine
Importation of Equine Animals Order 1979	Horses and other Equidae
Rabies (Importation of Dogs, Cats and other Mammals) Order 1974	Most warm-blooded mammals
Importation of Birds, Poultry and Hatching Eggs Order 1979	Live birds and eggs intended for incubation; any species

The controls imposed by these orders vary in detail from one to another but generally include the following requirements:

(a) An import licence must be obtained from MAFF before an animal is landed in Britain. The licence contains conditions which deal *inter alia* with (b)–(h). No licence is required for the import of up to two pet birds arriving in Great Britain accompanied by their owner who must sign an undertaking to HM Customs and Excise

(b) A quarantine period will be imposed: this is six months for most species under the 1974 Order and 35 days for birds; other quarantine periods are variable and are specified in the licence

(c) Transport to, and between, quarantine premises must be in a vehicle approved by MAFF
(d) Quarantine must be spent at premises authorised by MAFF. These may be permanent quarters such as quarantine kennels or catteries or may be specially approved premises, e.g. at a zoo or research establishment where, subject to proper precautions, the animals may be on view or in use during the quarantine period. In the case of birds, less stringent conditions of isolation are permitted instead of quarantine where importation is under a licence in category (i) (imports of up to 12 birds) or category (ii) (non-psittacine birds) or under the two pet birds exemption (see earlier)
(e) There must be regular veterinary inspection of quarantine quarters
(f) Permission must be obtained before an animal is moved out of quarantine premises
(g) Cleansing and disinfection of premises and equipment may be required, especially when an infected animal is found
(h) An infected animal which is landed in Britain may be disposed of by way of isolation, re-export or slaughter
(i) An import licence is likely to contain requirements for veterinary examination of the animal and health certification to be carried out before it leaves the country of departure

The rabies provisions impose additional requirements, e.g. vaccination against rabies for imported cats and dogs. An animal which has been taken, however temporarily, out of the British Isles and landed in another country is subject to quarantine on return. There are strict provisions to prevent animals on board boats in British harbours from coming into contact with animals ashore. Offences against the rabies legislation carry heavy penalties including destruction of the animal involved.

Special arrangements may be made for animals travelling to and from Great Britain for the purposes of sport, entertainment or breeding.

There are also orders which control the importation of ancillary items with which disease could enter the country. The orders and the material and animals to which they apply are listed below:

Importation of Embryos, Ova and Semen Order 1980	Embryos, ova and semen of any mammal except man; semen of poultry
Importation of Animal Products and Poultry Products Order 1980	Anything originating or made from a living or dead horse, ass, mule, zebra, swine, bovine animal or from living or dead domestic fowls, turkeys, geese, ducks, guinea-fowls,

	pigeons, pheasants and partridges and quail
Rabies Virus Order 1979	The lyssa virus of the family Rhabdoviridae (other than that contained in a medicinal product which may be imported under the Medicines Act 1968)
Importation of Animal Pathogens Order 1980	Any collection or culture of organisms or any derivative thereof (not contained in a medicinal product) which may cause disease in cattle, sheep, goats and all other ruminating animals, horses, swine and the birds listed for the 1980 Order earlier.
Importation of Bovine Semen Regulations 1984	Bovine semen

A licence is required to authorise the importation of material covered by these orders. (Cooper and Cooper, 1987). Additional powers include the seizure of material and the cleansing and disinfection of containers and vehicles. It is also an offence under the Rabies Virus Order to keep the virus or deliberately to introduce it into an animal. See also Cooper (1987).

Importation of Fish

The Diseases of Fish Act 1937 as amended by the Diseases of Fish Act 1983 forbids the importation of salmonid fish without a licence issued by the MAFF Fisheries Department. Any live freshwater fish, including those of the salmonid family and their eggs, which are to be imported into Britain must be consigned to a person who is licensed by MAFF to import them. The licence, set out in the Diseases of Fish Regulations 1984, may include provisions as to the disposal, transport, inspection, cleaning and disinfection of the fish or eggs and their containers, the means of transport and any other conditions to prevent the spread of disease. There are powers of seizure in respect of illegally imported fish.

The Import of Live Fish (England and Wales) Act 1980 and the (Import of Live Fish (Scotland) Act 1978 as amended by the Fisheries Act 1981 provide for the licensing of the importation, release or keeping of non-indigenous fish which may be inimical to freshwater fish, shellfish, salmon or their habitat. Licences for the importation of aquarium, usually tropical, fish have been freely available as such imports pose no threat to farmed fish stocks; however, this policy is to be reviewed in 1987.

Importation of Insects

The Bees Act 1980 and the Importation of Bees Order 1980 require that bees must be imported under licence. Other species of insects which constitute pests of plants may be imported only under the conditions of the Import and Export (Plant Health) (Great Britain) Order 1980, made under the Plant Health Act 1967. See Appendix 3 (Note 17).

Importation of Pest Species

The Destructive Imported Animals Act 1932 controls the importation of foreign animals which cause damage in Great Britain. Importation, except under MAFF licence, is prohibited by orders made in respect of musk rats and musquash (Musk Rats (Prohibition of Importing and Keeping) Order 1933), non-European rabbits (Non-Indigenous Rabbits (Prohibition of Importing and Keeping) Order 1954) and grey squirrels (Grey Squirrels (Prohibition of Importing and Keeping) Order 1937). The importation of hares is controlled by the Hares (Control of Importation) Order 1965 and that of mink and coypu under the Rabies (Importation of Dogs, Cats and other Mammals) Order 1974. Restriction is also imposed upon non-indigenous fish, molluscs and insects (see earlier).

Other Legislation on Importation

The requirements of MAFF regarding importation vary from time to time and it is important to ascertain the latest requirements when commencing importation procedures.

Importers of exotic species must consider the Endangered Species (Import and Export) Act 1976, which is administered by the Department of the Environment (see Chapter 7).

Customs duty is payable on any goods imported into Britain. Value added tax (VAT) may also be payable.

Animals imported for research or education may be eligible for relief from import duty (Common Customs Tariff) and, if they are supplied free of charge, from VAT also.

This exemption applies to all vertebrate, and some other, animals which have been specially prepared (e.g. purpose-bred) for laboratory use. They must be intended for use by public or private establishments principally engaged in education or scientific research. Private establishments must hold a letter of authorisation from the Home Office in order to obtain relief from import duty or VAT.

EXPORTATION

The Diseases of Animals (Export Health Certificates) Order 1985 requires export certificates for bovine animals, swine and fresh meat (including poultry) going to EEC countries. There are special orders relating to farm animals being sent to Canada and other countries which require quarantine before departure (Export Quarantine Stations Regulation Order 1973). A Department of Trade certificate may be required for the export of certain farm animals.

The Export of Horses (Protection) Order 1969 (and allied orders) and the Export of Animals (Protection) Order 1981 contain provisions to ensure proper standards of animal welfare for farm animals and horses being sent abroad.

Whatever the species involved, export procedures demand the completion of considerable documentation which is required by the importing country. Licences, health certificates, veterinary inspection, vaccination or diagnostic tests may be requested and these must be provided by MAFF officials. Such matters require careful planning and good timing; thus the guidance of official and commercial organisations involved in any shipment is invaluable. Information on current import requirements of foreign countries can be obtained from MAFF in some cases, otherwise from the appropriate diplomatic mission.

REFERENCES

Bell, J.C. (1981). Legal and ethical aspects of salmonellosis. *Veterinary Record* **109**, 300–304.

Bennett, G.H. (1976). The Zoonoses Order 1975. *Journal of the Royal Society of Health* **I**, 19–20.

Cooper, M.E. (1987). Legal considerations in the investigation of avian disease. Paper presented at the World Veterinary Congress, Montreal, August 1987.

Cooper, J.E. and Cooper, M.E. (1987). Diagnostic/screening specimens from overseas. Import permits. World Veterinary Congress, Montreal, August 1987.

Lowes, E. (1975). The Zoonoses Order 1975. *Veterinary Record* **97**, 32–33.

RECOMMENDED READING

See Chapter 1 for literature generally applicable, particularly Chapters II and VII of RCVS (1987).

Gregory, M. (1974). *Angling and the Law*, 2nd edn, with Supplement (1976). Charles Knight, London.

MAFF (1968). *Handbook of Orders Relating to Diseases of Animals*. Also Supplement No. 1 (1976), No. 2 (1978), No. 3 (1980). HMSO, London.

MAFF (1982). *Return of Proceedings under the Animal Health Act 1981, for the Year 1982*. HMSO, London.

MAFF (1983). *Animal Health 1982*. Report of The Chief Veterinary Officer, HMSO, London.

MAFF (1986). *Badgers and Bovine Tuberculosis* (by Dunnet, G.M., Jones, D.M. and McInerney, J.P.). HMSO, London.

Thomas, J.L. (1975). *Diseases of Animals Law*. Police Review Publishing, London (out of print).

MAFF produce guidance notes for those affected by animal health legislation. These are obtained from the relevant section of the Ministry (see Appendix).

6 Treatment and Care of Animals

May 22nd 1912
Prince (17 years old with one eye) receives his death blow from Spalding after being on Vets hands for seedy foot.
(From the "Horse Book" of Priday, Metford and Company Limited, Gloucester, flour millers)

Any person who has responsibility for an animal must be concerned for its health and should be prepared that, at some time or other, it may be necessary to give or obtain treatment for its illness or injury.

VETERINARY SURGERY

Right to Practise

In the UK the professional treatment of animals is regulated by the Veterinary Surgeons Act 1966. With certain exceptions, discussed later, only those people who are registered with the Royal College of Veterinary Surgeons may give veterinary treatment to animals. The rights and duties under the Act are described by Porter (1987) and the professional obligations of the veterinary surgeons are set out in RCVS (1987).

Those so registered include veterinary surgeons, who nowadays qualify by obtaining a veterinary degree from a UK university and being admitted to membership of the RCVS, and veterinary practitioners, included in the Supplementary Register, a closed list of those who are entitled to practise by virtue of their experience. Certain Commonwealth and South African veterinary degrees are recognised and the holders are eligible for registration. The Veterinary Surgeons Qualifications (EEC Recognition) Order 1980 provides for the recognition of qualifications acquired in EEC countries, and for the registration of their holders enabling them to practise in the UK on either a permanent or a temporary basis. Veterinary surgeons from the Commonwealth and other countries outside the EEC whose qualifications do not render them eligible for registration may sit an examination held by the RCVS and, if successful, will be entitled to membership of the College and to registration. For simplicity all such registered persons will be referred to as veterinary surgeons in this book, although "veterinarian" is more commonly used in other countries of the world.

Veterinary students are permitted to treat animals as part of their training in accordance with the restrictions imposed by the Veterinary Surgeons (Practice by Students) Regulations Order of Council 1981. A student may as part of his clinical training examine animals unsupervised, perform tests under direction, treat under supervision and operate under direct and continuous supervision and in accordance with the direction of a veterinary surgeon.

Animals

The restrictions on the right to treat animals relate only to those species which are covered by the Veterinary Surgeons Act which provides in section 27(1) that "animals includes birds and reptiles". This is usually interpreted to mean that fish and invertebrates are excluded from the ambit of the Act but it is not entirely certain whether amphibians are to be classed with reptiles or fish; marine mammals are considered to be subject to the Act. There is no restriction as to who may treat the excepted species. Nevertheless, any person, whether a veterinary surgeon, animal nurse or technician or unqualified person, who treats an animal is subject to other legislation such as the Protection of Animals Act 1911 and the Protection of Animals (Anaesthetics) Acts 1954 and 1964, which require the humane treatment of any species while in captivity (see Chapter 3), and the Medicines Act 1968 and Misuse of Drugs Act 1971, which place restrictions on the supply and administration of many drugs (see later in this chapter).

The majority of familiar animals, however, are covered by the Veterinary Surgeons Act and must be treated by a veterinary surgeon unless the circumstances fall into one of the exemptions provided by the Act.

Veterinary Surgery

The Veterinary Surgeons Act specifies the nature of work which is restricted to veterinary surgeons. It is defined by section 27(1) as "the art and science of veterinary surgery and medicine . . . to include —

(a) the diagnosis of diseases in, and injuries to, animals including tests performed on animals for diagnostic purposes;
(b) the giving of advice based upon such diagnosis;

(c) the medical or surgical treatment of animals; and
(d) the performance of surgical operations on animals.''

There are several exceptions to this requirement whereby the non-veterinarian may give treatment to animals. These are set out in sections 19(3) and 19(4) and Schedule 3 of the Act as supplemented by the Veterinary Surgery (Exemptions) Orders 1962, 1973 and 1982 and amended by the Veterinary Surgeons Act 1966 (Schedule 3 Amendment) Order 1982. At the time of writing (June 1987) certain changes to Schedule 3 are envisaged.

Treatment by Lay Persons

There are general exemptions from the Act, listed in Schedule 3, for the following:

(a) Any treatment given to an animal by its owner or his employee, by a member of the owner's household or his employee.

It is worth noting that the Act refers to 'any treatment' which, it might be argued, refers to item (c) in the definition of veterinary surgery and does not extend to other aspects of the definition such as diagnosis and surgical operations. In any situation, a non-veterinarian treating animals should be aware that the Protection of Animals Acts 1911–1964 (see Chapter 3) are applicable

(b) First aid measures taken by any person in an emergency to save life or relieve pain

(c) Anything, except a laparotomy, done other than for reward to an animal used in agriculture by its owner or a person employed or engaged in caring for such animals

(d) The castration of animals up to certain prescribed ages, the docking of lambs' tails and the docking of a dog's tail or the removal of its dew claws before its eyes are open

These must be carried out by a person who is at least 18 years old (or 17 if participating in animal husbandry training under the direct personal supervision of a veterinary surgeon or at a recognised institution under the direct personal supervision of an appointed instructor)

These exceptions are subject to an overriding provision that certain procedures must, nevertheless, be carried out by a veterinary surgeon. These restrict castration by the non-veterinarian to very young dogs, cats and farm animals; they require certain procedures in animal husbandry to be performed by a veterinary surgeon, as well as the removal of deer antlers in velvet, except in an emergency (Veterinary Surgeons Act 1966 (Schedule 3 Amendment) Order 1982).

Certain non-veterinarians may perform aspects of veterinary surgery:

(a) A person carrying out any procedure duly authorised under the Animals (Scientific Procedures) Act 1986 (see Chapter 4)
(b) Doctors and dental surgeons carrying out treatment, tests and operations, and persons giving treatment by physiotherapy to an animal, at the request of a veterinary surgeon; doctors may remove organs and tissues from an animal for the treatment of human beings
(c) Those authorised to take blood samples in farm animals and poultry under the Veterinary Surgery (Blood Sampling) Order 1983

The issue of veterinary and lay treatment has given rise to a number of questions to which solutions can at least be argued, even if there is little case law to give a conclusive answer.

It should be emphasised that in the UK (except in relation to Schedule 3, item (c) (see earlier)) it is irrelevant whether payment is received for treatment; the non-veterinarian, and it may well be an animal welfare organisation such as a rescue centre, cannot give veterinary treatment to other people's animals by means of non-veterinarians even if this is on a non-paying basis. In certain other countries, notably the USA, lay people may treat animals provided that no charge is made and the restriction of treatment to veterinarians applies only to domestic species.

Non-veterinarians may carry out laboratory tests on material taken from animals by veterinary surgeons — a service often offered by specialist companies — but care must be taken that any diagnosis or advice which is based on the results of such tests is given by a veterinary surgeon.

There would seem to be no objection in the Act to simply stating a treatment for a given condition provided that no diagnosis is offered by the adviser. This would cover the natural inclination of people to discuss remedies.

A popular new subject is the treatment of behavioural problems in animals and a very fine line has to be drawn between this, which is often carried out by non-veterinarians, and veterinary treatment which may also be required in order to deal with the problem fully.

Difficulties may arise in the care of sick or injured free-living wild animals found by non-veterinarians. The latter may, of course, give emergency first aid in the field but unless the animal is taken into captivity (cf. temporary restraint, see below) for further attention (in the case of protected species, in accordance with the provisions of wildlife conservation legislation (see Chapter 7)) then the restrictions of the Veterinary Surgeons Act must be observed. In civil law a wild animal only becomes a person's property when it has been reduced into his ownership by his exercising some form of control over it (see Chapter 1). As the owner, he could rightly give it veterinary attention or obtain the services of a veterinary surgeon. If he seeks the help of a non-veterinarian, e.g. an animal rescue centre, ownership must be given to the centre if it is to provide the necessary treatment.

It has been held that the brief and temporary restraint of a wild animal does not render it captive (Rowley *v.* Murphy [1964]) (but see Chapter 3).

The foregoing reasons thought must be given to the Act when wildlife is temporarily caught for scientific purposes, especially with a view to applying superficial monitoring devices or using painless methods of identification. If no diagnosis, treatment or surgical intervention is involved the activity is not governed by the Veterinary Surgeons Act. If veterinary procedures are involved, going beyond first aid, they should be performed by a veterinary surgeon. This does not apply to regulated procedures authorised by the Animals (Scientific Procedures) Act 1986 (see Chapter 4).

It is sometimes asked whether anaesthesia and sedation fall within the definition of "veterinary surgery". In the absence of a judicial decision it is taken as a working rule that they are classified according to the procedure to which they relate. Thus, anaesthesia prior to veterinary treatment is covered by the Veterinary Surgeons Act, whereas that used solely for management or handling is not considered veterinary surgery and the anaesthesia may be carried out by a non-veterinarian.

If anaesthesia is used prior to an experimental or other scientific procedure carried out under the Animals (Scientific Procedures) Act 1986 it constitutes a regulated procedure and the person administering it need not be a veterinarian provided that he is duly authorised under that Act. A veterinary surgeon carrying out anaesthesia as a regulated procedure must also be authorised under the 1986 Act (see Chapter 4).

An organisation which owns its animals, such as a research institution, may use its own non-veterinary staff to treat the animals or, of course, to give first aid. Care must be taken not to exceed these exceptions to the Act nor to allow treatment by lay staff of animals which, although part of a research programme, do not belong to the institution because they were purchased by a research worker out of his personal grant.

Veterinary nurses, trained and examined under the RCVS Veterinary Nursing Bye-laws, do not have any special authority and are subject to the same restrictions in respect of veterinary surgery as other lay people (Anon, 1986).

The practice of veterinary surgery by members of the profession and procedures carried out by non-veterinarians under the exceptions to the Act must be considered in conjunction with other aspects of law, in particular the need to obtain the owner's consent to treatment of an animal, to avoid actions for trespass and to conform with the Protection of Animals Acts 1911–1964 (see Chapter 3) and the medicines legislation (see later).

MEDICINES

General

The production, importation, supply, sale and administration of medicinal products are controlled by the Medicines Act 1968. The Misuse of Drugs Act 1971 imposes special restrictions in respect of "controlled drugs". The legislation applies to medicines used for human as well as veterinary treatment although reference here will only be to the latter. It is not contrary to the medicines legislation to use drugs produced and licensed for humans in animals although a veterinarian should inform a client if he proposes to use either a product not licensed primarily for animal use or a product for a species not mentioned on the data sheet (Cooper, 1985).

Medicines Act 1968

The importation (except for UK-licensed drugs from EEC countries permitted by the Medicines (Veterinary Drugs) (Exemption from Licences (Importation) Order 1986) and production (including the manufacture, development and testing) of a veterinary medicinal product must be carried out under a product licence.

The sale and supply of medicinal products are regulated according to category. Those on the General Sale List (e.g. certain anthelmintics and flea powders) may be obtained without a prescription and some of them may be sold elsewhere than in a pharmacy. Pharmacy drugs (marked P) are not on prescription but are obtainable only from a registered pharmacy or a veterinary surgeon. There is also a special category of products which would otherwise have been on the P list and which relates only to veterinary medicines. This is the PML list of medicines available not only from veterinarians and pharmacists but from agricultural merchants without prescription provided that the buyer is a farmer or other person who has in his charge or maintains animals for business purposes. The incorporation of PML or POM products in animal feeds must be authorised by the written direction of a veterinary surgeon (BVA/RCVS/ADAS, 1986). Some horse worming drugs may be sold by saddlers. Codes of conduct also relate to PML and saddlers' list drugs (Anon, 1985; MAFF, 1985a, 1985b).

Drugs (marked POM) which are on the Prescription Only List of the Medicines (Veterinary Drugs) (Prescription Only) Order 1985 may be sold or supplied by a pharmacist on the authority of a prescription issued by a veterinary surgeon (s. 58(2)(a)). The prescription must state that the drugs are to be supplied to animals under the prescribing veterinary surgeon's care. It should be noted that, although the Medicines Act permits a veterinary surgeon to dispense POM or PML medicines to clients, he may only do so if the products are for administration to an animal or herd under his care (s. 58(3)(b)).

The phrase "under his care" has received special attention in the Guide to Professional Conduct (Appendix 7) (RCVS, 1987) and requires that a veterinary surgeon should have some real, not nominal, responsibility for the health of the animals concerned and should be acquainted with them to the extent that he should have seen them recently or have sufficient personal knowledge of their state of health to be able to diagnose or prescribe properly for them. The application of this principle to free-ranging animals such as a herd of deer has been considered by Porter (1982).

Prescription only medicines may be supplied to universities or other institutions concerned with research or higher education. The establishment concerned must provide to the supplier an order signed by its principal or the head of department in charge of a specified course of research. The order must give the name of the institution, the quantity of medicine required and the purpose for which it is required (Medicines (Veterinary Drugs) (Prescription Only) Order 1985, Schedule 3, Part 1). The medicinal products may only be used for the purposes of the education or research with which the institution is concerned.

Veterinary prescription only medicines may only be administered by a veterinary surgeon or a person acting in accordance with the directions of a veterinary surgeon, whether they are required on prescription from a pharmacy, through a veterinary surgeon or by a scientific establishment (Medicines Act s. 58(2)(b)).

The phrase "in accordance with the directions" has not received any judicial attention but it is considered that a veterinary surgeon has discretion as to what degree of direction, from verbal guidance to actual oversight of its use, he should give to a client. It must, however, be appropriate for the circumstances since he could incur a degree of liability for negligent advice or supervision.

There are many other detailed aspects of the Medicines Act and relevant Orders which concern veterinary surgeons and pharmacists such as prescribing, storage, labelling and merchants' list medicines. These have already been extensively considered by ABPI (1982), Knifton and Edwards (1987), Harrison (1987) and Pearce (1984).

Misuse of Drugs Act 1971

The Misuse of Drugs Act 1971 and the Misuse of Drugs Regulations 1985 impose restrictions upon "controlled drugs" (such as etorphine and other morphine derivatives). These provisions apply in addition to those relating to the prescription only medicines under the Medicines Act.

The possession, production and supply of controlled drugs are restricted and are *prima facie* illegal (s. 5) although exceptions are made by the Regulations. The controlled drugs most commonly administered to animals are contained in Schedules 2 and 3 of the Regulations. Schedule 2 drugs held by veterinary surgeons must be kept in a locked safe, cabinet or room so as to prevent unauthorised access, or in a locked receptacle which can only be opened by him or a person authorised by him. Records of usage must be kept. Those listed in Schedule 3, including barbiturate drugs, are not at present subject to the requirements for record keeping or (in most cases) safe custody (Edwards and Knifton, 1984, 1986).

Veterinary surgeons are permitted to supply and possess controlled drugs when acting in their capacity as such. Supply must be to a person who may lawfully have that drug in his possession.

A person in charge of a laboratory the recognised activities of which include the conduct of scientific education or research and which is attached to a university, university college or a hospital maintained out of public funds or by a charity or other institution approved by the Home Secretary, when acting in his capacity as such, may possess or supply controlled drugs listed on Schedules 2 and 3 of the Regulations. Supply must be to a person legally entitled to have the drug in his possession.

In addition to the foregoing a person may possess a controlled drug in Schedule 3 for administration for veterinary purposes in accordance with the directions of a veterinary surgeon.

The term "veterinary purposes" must be examined briefly. Again, there is no definitive interpretation of these words. If the term is taken to mean the purposes of veterinary surgery as defined in the Veterinary Surgeons Act 1966 (s. 27(1)) then this would preclude the supply of controlled drugs for certain activities not falling within that Act (such as herd management or biological studies). However, if a broader view is taken of "veterinary purposes" and it is assumed to include anaesthesia of an animal then one must note that it does not accord with the approach to the Veterinary Surgeons Act whereby the use of anaesthetic or immobilising drugs is classified as veterinary surgery or not according to the purpose for which anaesthesia is induced (see earlier in this chapter).

The Misuse of Drugs Act section 13 enables the Home Secretary to restrict the right of a veterinary surgeon to prescribe, administer or supply controlled drugs if he considers him to have been prescribing them irresponsibly. Consequently veterinary surgeons must ensure that directions are given with care. There are no exact guidelines or judicial decisions on this point, and directions may be detailed or general, written or spoken. The veterinary surgeon should satisfy himself that the client is genuine and that the amount of drug supplied is reasonable in the circumstances — in effect, conscientious supervision in accordance with standards currently accepted by veterinary surgeons in general. The presence of the veterinary surgeon during administration is not essential — it is another matter of judgment.

The Veterinary Defence Society has produced an agreement which veterinary surgeons who prescribe Immobilon (which contains etorphine) are encouraged to ask their clients to sign. This includes requirements as to possession, administration, storage and records. It gives information on safe handling of the drug and care of the animals on which it is used.

FIREARMS

A number of provisions apply to the use of firearms for the management and treatment of animals. These include:

Firearms legislation restricting the right to use certain weapons
Medicines legislation controlling the possession, supply and administration of medicinal products used in darting techniques
Wildlife legislation controlling the killing, injury or taking of wildlife and the means of doing so (see Chapter 7)

The weapons most commonly used are:

Rifle	Crossbow
Shotgun	Blowpipe
Captive bolt pistol	Dart gun

Pole syringes are not subject to this legislation.

Firearms Act 1968

A firearms certificate must be obtained from the police to authorise the acquisition and possession of any firearm (including a prohibited weapon) other than a shotgun or air weapon (s. 1, s. 26).

A firearms certificate is required for a captive bolt pistol although a person licensed under the Slaughterhouses Act 1974 does not require a certificate to possess a slaughtering instrument in a slaughterhouse (s. 10).

A shotgun certificate must be obtained from the police in respect of shotguns (s. 2).

Any weapon (e.g. a dart gun, blowpipe or crossbow) using ammunition containing any "noxious thing", e.g. an immobilising drug, is a "prohibited weapon". A Home Office authority must be held by those who acquire or possess a prohibited weapon (s. 5).

Applications for firearms and shotgun certificates are made to the chief of police for the area in which the applicant resides and authority for a prohibited weapon must be obtained from the Home Office prior to obtaining a firearms certificate (s. 20).

It is an offence to have a firearm in a public place without lawful authority (s. 20) or to trespass with a firearm without lawful excuse (s.20). The appropriate certificate should always be carried with a firearm or shotgun since a constable may require its production and may seize and detain the weapon on failure to produce a certificate.

The drugs to be used in a prohibited weapon are subject to the Medicines Act 1968 and, in many cases, the Misuse of Drugs Act 1971 (see earlier in this chapter). The user may be a veterinary surgeon who is entitled to possess and administer them in the course of his work as a veterinary surgeon. In other cases the drugs must be supplied either by or on the prescription of a veterinary surgeon in respect of animals under his care. They may also be acquired by a scientific or educational institution (see Medicines in this chapter).

Possession of controlled drugs is restricted and if they are to be held by persons not falling into the categories above the latter must hold a licence from the Home Office.

The administration of prescription only drugs, including controlled drugs, must be by or in accordance with the instructions of a veterinary surgeon regardless of the authority to supply or possess them.

CASE

Rowley *v*. Murphy [1964] 1 All E.R. 50.

REFERENCES

ABPI (1982). *The Safe Storage of and Handling of Animal Medicines*. Association of the British Pharmaceutical Industry, London.

Anon (1985). Code of conduct for dispensing of POM's to farmers. *Veterinary Record* **116**, 597.

Anon (1986). Veterinary nurses, 25 years on, *Veterinary Record* **119**, 437–438.

BVA/RCVS/ADAS (1986). *The Veterinary Written Direction and Withdrawal*

Periods. British Veterinary Association, London; Royal College of Veterinary Surgeons, London.
Cooper, J.E. (1985). The exotic pet — an introduction. In *Manual of Exotic Pets* (Cooper J.E., Hutchison, M.F., Jackson, O.F. and Maurice, R.J. eds). British Small Animal Veterinary Association, Cheltenham.
Edwards, B.R. and Knifton, A. (1984). Barbiturates and certain other drugs to be controlled under the Misuse of Drugs Act. *Veterinary Record* **115**, 649–650.
Edwards, B.R. and Knifton, A. (1986). Forthcoming changes in misuse of drugs legislation. *Veterinary Record* **118**, 350.
Harrison, I.H. (1987). *The Law on Medicines*. MTP Press, Lancaster.
Knifton, A. and Edwards, B.R. (1987). Controlled drugs and medicinal products. In *Legislation Affecting the Veterinary Profession in the United Kingdom*, 5th edn. Royal College of Veterinary Surgeons, London.
MAFF (1985a). *Code of Practice for Merchants Selling or Supplying Veterinary Drugs*. Ministry of Agriculture, Fisheries and Food, Alnwick.
MAFF (1985b). *Code of Conduct for Saddlers Selling or Supplying Horse Wormers*. Ministry of Agriculture, Fisheries and Food, Alnwick.
Pearce, M.E. (1984). *Medicines and Poisons Guide*, 4th edn. Pharmaceutical Press, London.
Porter, A.R.W. (1982). Drugs for use in dart guns. *Publication of the Veterinary Deer Society* **1**(2), 2–4 (also in RCVS (1987)).
Porter, A.R.W. (1987). Practice of veterinary medicine and surgery. In *Legislation Affecting the Veterinary Profession in the United Kingdom*, 5th edn. Royal College of Veterinary Surgeons, London.
RCVS (1987). *Guide to Professional Conduct*. Royal College of Veterinary Surgeons, London.
RDS (1974). *Guidance Notes on the Law Relating to Experiments on Animals*. Research Defence Society, London.

RECOMMENDED READING

See Chapter 1 for literature generally applicable.

Anon (1982). *Medicines and Poisons Guide*. Pharmaceutical Press, London.
BFSS (1975). *The Gun Code*. British Field Sports Society, London.
BVA (1981). Projector weapons: RCVS code of practice. In *Health and Safety at Work Act. A Guide for Veterinary Practices*. British Veterinary Association Publications, London.
Clarke, P.J. and Ellis, J.W. (1981). *The Law Relating to Firearms*. Butterworth, London.
Cassell, D. (1987). Veterinary and other services. In *The Horse and the Law*. David & Charles, Newton Abbot.
Dale, J.R. and Appelbe, G.E. (1983). *Pharmacy Law and Ethics*, 3rd edn. Pharmaceutical Press, London.
Home Office (1978). *The Use and Safe-keeping of Tranquillising Weapons*. Home Office, London, Scottish Home and Health Department, Edinburgh.
Pattison, I. (1984). *The British Veterinary Profession 1791–1948*. Allen, London.
Porter, A.R.W. (1974). Law, ethics and morality: the practising veterinary surgeon and the Medicines Act. *Veterinary Record* **95**, 407–411.
Sandys-Winsch, G. (1979). *Gun Law*. Shaw, London.

7 Conservation

Whilſt the ravenous Beaſts of Prey were ſo numerous in the Royal Woods, as to prevent the Increaſe of the Beaſts of delicious Taſte for the Table, the Kings gave free Liberty to the Nobility and Gentry to hunt in their Woods; but in Edgar's Time, the Breed of ravenous Beaſts being much leſſened, he having an elegant Taſte prohibited Hunting his Deer, and appointed Officers to preſerve all Game of the Table, in his Woods, who ſo rigorouſly put in Execution their Orders, that the Nobility and Gentry were prevented of taking their Diverſion, and their Tenants of their reſpective Rights.

(Stat.II Hen.7. c.17, 1496)

He who brings any Eyeſs-Hawk from beyond the Sea, ſhall have a Certificate under the Cuſtomer's Seal where he lands; or if out of Scotland, then under the Seal of the Lord Warden or his Lieutenant, certifying that ſhe is a foreign Hawk, upon Pain of forfeiting the Hawk.

(Nelson, W. "The Laws Concerning Game", 5th edn. T. Waller, London, 1753)

Legal protection in Great Britain for non-domesticated species falls into four categories:

(1) Conservation: the Wildlife and Countryside Act 1981 Part I and the Badgers Act 1973 provide protection for wild birds and certain other species by imposing restrictions on killing, taking, possession of and trading in such species. Some protection for their habitat is also provided, both in Part I with reference to protected species and in Part II in respect of specific areas of countryside.

(2) Close season protection: some wild animals (seals, deer, some birds and fish) which are used in sport are protected during their breeding season and there are limitations as to permitted methods which may be used for taking or killing them.

(3) Control of pests: many wild species (e.g. foxes, weasels and many insects) are not protected in Great Britain. However, they and others, such as rats or grey squirrels, may be subject to control measures because they constitute pests (UFAW, 1985). Every species protected under (1) and (2) above may, in certain circumstances, be killed to pre-

vent serious damage to crops or other property. The import, keeping and release of non-indigenous species into the wild in Great Britain (including some already established there (see later)) is prohibited.

(4) Trade control: there are restrictions on the sale of species protected by the conservation legislation and controls on the sale of dead game (in conjunction with poaching controls); in addition, the Endangered Species (Import and Export) Act 1976 (as amended) and EEC Regulation 3626/82 (as amended) contain extensive provisions as to the trade in non-domesticated species.

Even when species are not protected by legislation, voluntary codes may be drawn up to promote their conservation, as in the case of insect collecting (JCCBI, undated). See Appendix 3 (Note 18).

WILDLIFE AND COUNTRYSIDE ACT 1981

The conservation of free-living endangered species (other than birds where there has been long-standing legislation, or in specific nature reserves) first appeared in Britain in the 1970s. Until that time such creatures had no protection, although their captive relatives were, and still are, protected from cruelty by the Protection of Animals Act 1911 and allied legislation (see Chapter 3). Habitat protection, apart from that provided by existing nature reserves and other limited provisions, was introduced by the Wildlife and Countryside Act 1981 which also revised the previous protection for wild birds and other creatures.

The Wildlife and Countryside Act 1981 (WCA) came into force on various dates but the parts relating to wildlife were fully effective from 28 September 1982. The Act implements Great Britain's obligations under the EEC Directive on the Conservation of Wild Birds and the Council of Europe Convention on the Conservation of European Wildlife and Natural Habitats (see Chapter 9).

Conservation of Birds

The WCA (which replaced the Protection of Birds Acts 1954–1961) provides protection, to a greater or lesser extent, to wild birds in Great Britain.

Protected birds

The Act applies to "any bird of a kind which is ordinarily resident in or is a visitor to Great Britain in a wild state". However, it applies to game birds (pheasants, partridge, red grouse, black game (black grouse) or ptarmigan) only in respect of prohibited methods of trapping (s. 5) and does not apply at all to poultry (domestic fowls, geese, ducks, guinea-fowls, pigeons, quails and turkeys) (s. 27(1)).

The definition is taken to apply to any bird which is on the "British List" produced by the British Ornithologists' Union. It may, therefore, encompass species not regularly found in Great Britain. It also includes British birds which are in captivity, were formerly in captivity but have escaped or been released and, *prima facie*, have been bred in captivity.

Section 1(6) provides, however, that the basic protection of that section (see later) does not apply to "any bird which is shown to have been bred in captivity". It should be noted that this wording puts the burden of proof upon the person alleging that a bird is captive-bred — most commonly the defendant to a prosecution under that section. In addition, section 27(2) provides that for a bird to be treated as captive-bred its parents must have been lawfully in captivity when the egg was laid. This second factor can be proved by production of the relevant documents when a bird has been imported or taken under licence. When birds have been obtained by purchase, inheritance or under the exemption for sick or injured birds (see later) documentation may be less readily available; in the case of captive breeding itself it may be necessary to trace ancestry back to the original stock. Where these were obtained before the 1981 Act or import restrictions were imposed it may be difficult to provide conclusive evidence of origin or captive breeding. Even with a present-day breeding it is not easy to prove that a particular egg was laid by a given pair of birds. The importance of regular observation, accurate record keeping and, in the case of "monitored species" (peregrine, merlin, goshawk, golden eagle, gyr falcon and their hybrids), the need to have the breeding activity confirmed by DOE inspectors (DOE, 1987) cannot be overemphasised. Section 1 does not apply to captive-bred or foreign birds; other parts of the Act are applicable to them. Thus, although it is permissible for a person to recapture an escaped captive-bred or foreign bird, he must not use any of the methods prohibited by section 5 without a licence authorising their use. A wild-bred British bird which escapes from captivity cannot be recaptured without a further licence to take it in addition to authority to use prohibited methods of capture.

Protection

The basic protection for wild birds in WCA section 1 provides that it is an offence to:

s. 1(1)	Kill Injure Take	Any WILD BIRD

Section	Act	Object
	Take Damage Destroy	The NEST, while in use or being built, of any wild bird
	Take Destroy	An EGG of any wild bird
s. 1(2)	Have in one's possession or control	Any LIVE Any DEAD Any PART of a Any DERIVATIVE of a Any EGG of a Any PART of an egg of a } Wild bird
s. 1(3)	Unless it is proved that: (a) Bird or egg was NOT TAKEN or (b) Had been killed or taken LEGALLY or (c) Had been SOLD LEGALLY	
s. 1(5)	Disturb	A Schedule 1 bird while it is building its nest or near a nest containing young or its dependent young
s. 6(1), (2)	Sell	Any live wild bird unless in Schedule 3, Part I, any dead wild bird except Schedule 3 Part II or III birds unless seller is registered under section 6(3)

The Acts prohibited by sections 1(1) and 1(3) must have been committed intentionally to constitute an offence.

Illegal possession (s. 1(2)) arises independently of any intention. Thus in Kirkland *v*. Robinson (1986) it was held that section 1(2) created an offence of strict liability. In other words, the reasonable belief that one possesses wild birds legally, or even the absence of knowledge that one's possession is illegal, is no defence to a prosecution under this section. Lawful possession is a matter of fact which must be proved by the defendant and can be done by showing that a bird has been captive bred, imported under licence or taken from the wild under licence. The case involved goshawks which the prosecution proved not to have been captive bred, contrary to the defendant's claim. Although not mentioned in the case report, another ground for legal possession is the keeping of a sick or injured wild bird pending its recovery and release (see later).

These provisions for protection of wild birds are applied in varying degrees according to their status. This can be summarised as follows:

Schedule 1 Part I	Rarer birds, e.g. diurnal birds of prey, some owls, many smaller birds	Full protection + special penalty £2000 (s. 1(4))
Part II	e.g. pintail	As above, during close season only (s. 1(7)) (see Table 1)
Schedule 2 Part I	Certain wildfowl and game birds	May be killed or taken outside their close season (see section 2(1)) except during a period of special protection (s. 2(6)) (see Table 1).
Part II	Pest species, e.g. crow, house sparrow, feral pigeon	May be killed or taken. Nest may be damaged. Nest or egg may be destroyed at any time by an authorised person (s. 2(2))
Schedule 3 Part I	Certain captive-bred species, e.g. blackbird, barn owl, starling, goldfinch	Sale of live birds permitted at any time provided they were bred in captivity and ringed or marked (s. 6(1), (5))
Part II	Feral and wood pigeon	May be sold dead at all times (s. 6(2), (6))
Part III	Certain game birds and wildfowl	May be sold dead 1 September–28 February (s. 6(2), (6)) (see Table 1)
Section 6(2)	Birds not in Schedule 3, Parts II or III	Vendor must be registered with DOE
Schedule 4	British or foreign wild- or captive-bred diurnal birds of prey (except owls, vultures); many rare birds	If kept in CAPTIVITY most must be RINGED and REGISTERED with DOE

In addition to the protection afforded by section 1 of the WCA, the DOE has power, with the consent of the landowners or occupiers affected, to make an order protecting a specified area in which all wild birds (except pest species) have the full protection of section 1 and the special penalty; in addition an order may restrict access to the area (s. 3).

There are various exceptions (s. 4) to the provisions of sections 1 and 3; in the case of (b) to (d) below, sections 4(2) and (3) do not preclude prosecution but provide a defence to action taken for certain purposes which would otherwise be a breach of those sections. The exceptions are as follows:

(a) Anything done in compliance with MAFF requirements under section 98 of the Agriculture Act 1947 or Agriculture (Scotland) Act 1948 relating to pest clearance, or under the Animal Health Act 1981 in respect of disease control, particularly in wild animals (although a licence is required in respect of Schedule 1 birds) (s. 4(1))
(b) The killing or injuring of a wild bird (except Schedule 1 birds for which a licence must be obtained) by an authorised person for the purpose of protecting public health or safety, preventing the spread of disease or serious damage to livestock, their food, to crops, vegetables, fruit, timber and fisheries (s. 4(3)). An "authorised person" includes the owner or occupier of the land involved or a person authorised by them and also a person authorised by certain bodies (s. 27(a)) (see Licensing).
(c) Anything which would be an offence under sections 1 or 3 but which could not reasonably have been avoided. Such an incident must have arisen in the course of lawful activity (s. 4(2)(a)). Thus, to kill a bird accidentally while driving a car legally is not an offence; however, to kill a bird while speeding or trespassing would be in breach of sections 1 or 3.
(d) It is permitted to kill any wild bird which is so badly disabled that there is no reasonable likelihood of its recovery. Similarly, a sick or injured bird may be taken solely for the purpose of tending it and returning it to the wild when it has recovered (s. 4(2)(a) and (b)). In both circumstances the damage to the bird must not have been caused unlawfully by the person killing or taking it. Some birds never improve sufficiently to return to the wild. In such a case it may be wise to get a written veterinary opinion to support the reasons for keeping the bird. Schedule 4 birds must be ringed and registered (see later). If, when releasing a bird, there is doubt as to its ability to survive in the wild, attention should be given to the Abandonment of Animals Act 1960 (see Chapter 3).

Schedule 4 birds in captivity

The WCA section 7 requires that any wild birds listed in Schedule 4, when kept in captivity, must be registered with the DOE in accordance with the Wildlife and Countryside (Registration and Ringing of Certain Captive Birds) Regulations 1982, and most of these species must be ringed (DOE, 1983). Birds in Schedule 4 include all diurnal birds of prey (not owls), whether British or foreign species (except old world vultures), and many other rare, threatened or localised species such as the crossbills, the fieldfare, firecrest, redwing, spoonbill and some warblers. The provisions apply to both wild- and captive-bred specimens.

Disabled Schedule 4 birds which are taken from the wild must be ringed and registered unless exempted by general or individual licence. There is an open general licence (OGL) permitting veterinary surgeons to keep disabled wild birds for up to six weeks for the purpose of giving them veterinary treatment. "Licensed rehabilitation keepers" (LRKs) may keep Schedule 4 birds unlicensed for the same length of time. Thereafter, birds must be ringed and registered with the DOE and a fee must be paid, although LRKs pay a reduced fee. Both veterinary surgeons and LRKs must keep records of the registrable birds which they keep and, in addition, the premises of LRKs are subject to inspection.

The effect of section 7 is that all birds of prey and most other birds listed in Schedule 4 which are in captivity should bear a numbered ring (either a cable-tie or, for most birds bred in captivity in Great Britain from 1983, a closed ring) issued by the DOE. Any captive bird without a ring must be one held under the disabled bird OGLs or have specific authorisation.

The DOE must be notified when a registered bird is moved to a new address, is sold or otherwise disposed of, dies, escapes or is released to the wild or is exported. When there is a transfer to a new owner the bird must be re-registered. Re-registration is also required if a ring is removed, lost or becomes illegible.

Section 6 of the WCA makes illegal the sale of Schedule 4 birds, except when authorised by a licence. OGLs have been issued by the DOE permitting the sale of certain species of captive-bred wildfowl and birds of prey.

Prohibited methods of killing or taking wild birds

It is illegal (s. 5(1)(a)) to set in position certain articles (such as traps, snares, electrical devices or poisons) if they are "of such a nature and . . . so placed" as to be calculated to cause bodily injury to a wild bird coming into contact with them. This provision applies to articles used to take or kill any bird (whether or not a protected species, e.g. a parrot to be recaptured). "Calculated" is usually interpreted broadly to include "likely" to cause harm.

The use of such items, regardless of their nature or any intention to cause injury, to take or kill wild birds is also an offence (s. 5(2)(b)). Additional illegal items under this provision include nets, baited-boards, bird lime or similar substances. Indeed, simply being in possession of these and any other item is illegal if they are intended for the purpose of committing an offence under Part I of the Act (s. 18(2)).

Other articles which may not be used to kill or take wild birds include cross-bows, explosives, certain firearms, gas, smoke and chemical wetting agents.

Vehicles must not be used in immediate pursuit of a wild bird; certain forms of decoy are illegal, including sound recordings and live animals which are under restraint or physically incapacitated.

Licences may be issued to permit the use of prohibited methods for certain purposes. Section 5(4) specifically permits the placing of traps etc. to catch unprotected creatures for purposes such as public health, agriculture, forestry, fisheries and nature conservation, provided that all reasonable precautions are taken to prevent injury to wild birds.

Cage traps or nets may be used to take Schedule 2 Part II (pest) birds by an authorised person or to take game birds for breeding; duck decoys in use since 4 June 1954 may continue to be operated.

Miscellaneous provisions

Section 6(3) forbids the exhibition of live, wild, British birds for competition, or at the premises where a competition is being held, unless they are listed in Schedule 3 Part I and were bred in captivity and close ringed (see earlier). This provision does not apply to non-competitive exhibitions or to foreign species of bird except that it is a requirement that both parents of any bird exhibited must have been in captivity when it was bred.

Section 8 requires, with some exceptions, birds (wild or otherwise) to be kept in cages large enough to permit them to stretch their wings fully. This has been discussed in Chapter 3 because it is primarily a welfare provision. Bird shoots in which birds are liberated for immediate shooting are illegal.

Protection of Wild Animals (Other than Birds)

The WCA section 9 provides protection for the wild animals listed in Schedule 5 to the Act. Some species, e.g. red squirrel, bats, the common dolphin, some reptiles, amphibians and insects, have full protection; some reptiles and amphibians are subject only to the restrictions on sale, e.g. the adder, grass snake and common frog and toad (Cooper, 1986; Langton, 1986).

A wild animal is defined (s. 27(1)) as "any animal (other than a bird) which is or (before it was taken or killed) was living wild". There is a presumption that an animal is a wild one and so the burden of proof falls on the person alleging captive breeding.

The provisions follow those relating to wild birds in that it is an offence to

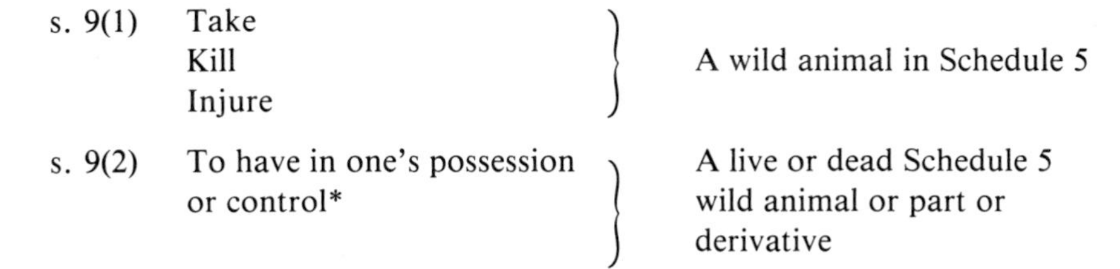

s. 9(1)	Take Kill Injure	A wild animal in Schedule 5
s. 9(2)	To have in one's possession or control*	A live or dead Schedule 5 wild animal or part or derivative

s. 9(5)	Sell Advertise for sale	A live or dead Schedule 5 wild animal or part or derivative
s. 9(4)	Damage Destroy Obstruct access to**	Any structure or place used for shelter or protection
	Disturb**	The use of such place

s. 9(3) *Unless it is proved not to have been taken or killed, nor taken, killed or sold in contravention of WCA Part I

s. 10(2) **Unless in a dwelling-house. In respect of bats, the NCC must first be notified unless it relates to the living area of a house

See also Appendix 3 (Note 19).

As in the case of wild birds (see earlier), intention is an essential element of an offence under sections 9(1) and 9(4).

Activities which may be carried out despite the foregoing provisions are comparable to those in section 4 relating to wild birds, together with the additional provisions in respect of bats mentioned above (s. 10(1)(3)(4)(6)).

Prohibited methods

These follow the provisions for birds (in WCA section 5) and are applicable to Schedule 6 species. In addition, the setting of self-locking snares likely to injure wild animals (not merely protected ones) is illegal as is the use for the purpose of killing or taking any wild animal (regardless of intention or likelihood of injury) of self-locking snares, bows, crossbows, explosives or live mammals or birds as decoys (s. 11(1)). See Appendix 3 (Notes 19 and 20).

Thus, while it is illegal to set any trap, snare, electrical device or poison which may injure a Schedule 6 animal or to use any such method to take or kill these protected species (s. 11(2)), it is permissible to do so in respect of unprotected species provided that none of the methods (such as poison) is forbidden by section 11(1) or by other legislation (see Chapter 3).

When unprotected species are being trapped, care must be taken that Schedule 6 species are not caught or harmed. Indeed, when setting traps or other devices to take permitted wild animals in the interests of public health, agriculture, forestry, fisheries or nature conservation it is a defence to show that all reasonable precautions were taken to prevent injury to Schedule 6 animals (s. 11(6)). If snares which are likely to injure wild animals are used they must be inspected at least once a day.

Licences must be obtained to take or kill Schedule 6 wild animals or to use prohibited methods. An OGL has been issued by the NCC to permit the trapping of shrews for identification and immediate release by members of scientific, educational or conservation bodies.

Licensing

These provisions apply to both birds and other creatures protected by the WCA. Most of the activities which are *prima facie* forbidden by the Act and do not fall within the exceptions already mentioned can be authorised by a licence issued by MAFF, the Nature Conservancy Council or the DOE (s. 16(9)). Licences will be issued under section 16 to possess, kill or take and to use otherwise forbidden methods of doing so for scientific and educational purposes, for ringing and marking, for conservation, for protecting animal collections, for falconry and aviculture and for photography. Licences may be issued for the taxidermy of birds only; the effect of this is that while a licence may be issued specifically to take birds from the wild for taxidermy, only protected animals acquired in conformity with sections 9 and 10 of the Act may be used for taxidermy.

A licence must generally be obtained to kill wild birds or Schedule 5 wild animals, contrary to sections 1 and 9 respectively, although this principle does not apply to:

(a) Birds to which the Act does not apply (poultry and foreign, captive-bred and game birds)
(b) Schedule 2 Part I birds outside the close seasons and Schedule 2 Part II birds
(c) The killing of any other bird not on Schedule 1 by an authorised person for the purposes of preserving public health or public or air safety and the prevention of disease or serious damage to livestock, crops and similar property (s. 4(3))
(d) Wild (non-avian) animals other than those on Schedule 5
(e) The killing of a Schedule 5 species in an emergency to prevent serious damage to livestock, crops and similar property (s. 10(4)(6))

It is advisable to obtain licences in these circumstances since the provisions in (c) and (e) only provide a defence to, and do not preclude, a prosecution and possible conviction under section 1. A licence must always be obtained for Schedule 1 birds (s. 4(3)) and must be applied for in respect of Schedule 5 species if it is reasonably practicable to do so (s. 10(6)(a)).

Release or Escape of Non-indigenous Species

A novel aspect of the Act is the provision of section 14 forbidding the release or escape into the wild of non-indigenous species of wild animal (defined as being "of a kind which is not ordinarily resident in Great Britain in a wild state''), including the species already established in Britain which are listed in Schedule 9, such as the grey squirrel, budgerigar, African clawed toad and Canada goose (NCC, 1979). No offence is committed if all due care was taken to prevent an escape or release.

This provision has led to some problems in practice and the DOE is reported to have provided opinions upon two issues, subject, of course, to judicial interpretation. The DOE consider that it is a violation of section 14 to allow aviary-nesting foreign birds (such as softbills) to fly in and out of an aviary while feeding their young in order to obtain sufficient live food (Low, 1985). The keeping of foreign frogs in an escape-proof domestic garden was considered not to constitute a release into the wild (s.4) (Beebee, 1983). The limitations of section 14 for species which escape readily into the wild is illustrated by Marren (1986) in respect of the American signal crayfish. As a consequence, the import, keeping and release of certain non-indigenous fish, crustacea or molluscs or their eggs is controlled by the Import of Live Fish (England and Wales) Act 1980; similar provisions are contained in the Import of Live Fish (Scotland) Act 1978.

A code of practice to control the introduction of insect populations for conservation and scientific purposes has been produced by the Joint Committee for the Conservation of British Insects (JCCBI, 1986).

Penalties

The Act (by section 21) imposes penalties for offences committed against its provisions; these are mainly fines ranging from maxima of £200 to £2000, the latter relating to special penalties in respect of a number of offences, particularly those involving trapping or Schedule I birds. For conviction in the Crown Court an unlimited fine may be imposed. Fines may be imposed cumulatively in respect of each animal or item involved and animals, articles or vehicles used in the commission of an offence may be forfeited.

A court is obliged to order the forfeiture of a bird, nest, egg, other animal or object in respect of which an offence has been committed. The right to possess Schedule 4 birds or to be registered in respect of Schedule 3 Part II and III birds is automatically lost for five years on conviction for an offence incurring a special penalty or for three years in respect of other WCA Part I offences relating to the protection of birds or other animals or any offence (under other legislation) involving their ill treatment (ss. 6(8) and 7(3)). The effect of these mandatory provisions is that prosecutions under the Act are more vigorously defended than under former legislation.

Enforcement

The DOE is responsible for enforcement of the Act, through its Chief Wildlife Inspector. Voluntary inspectors are also appointed and persons authorised by the DOE have a right of entry (at reasonable times and on production of his authority) to inspect premises:

(a) Where wild birds are kept by a person registered under section 8(7) in respect of Schedule 3 Part II or III birds, for the purpose of ascertaining whether or not an offence has been committed
(b) Where Schedule 4 birds are kept (s. 7(b))
(c) In respect of escapes and releases of foreign species (s. 14(5))

The Act permits the issue of warrants by a magistrates' court to authorise a police constable to enter and search premises for evidence of the offences created by the Act. When a constable has reasonable cause to suspect the presence of evidence of the commission of an offence under the WCA, he may, without a warrant, stop and search a person, seize and detain evidence in his use or possession and arrest him if he fails to give his name and address (s. 19) (NCA, undated).

Badgers Act 1973

The Badgers Act (as amended by the WCA Schedule 7 and the Wildlife and Countryside (Amendment) Act 1985) affords protection to badgers in terms very similar to those given to Schedule 5 animals under the WCA. In addition, it makes illegal various forms of cruelty to badgers including ill treatment, the use of badger tongs, badger-digging (except under licence) and the use of weapons other than specified calibre rifles. The problems of enforcing these provisions have been discussed by Harris (1986).

CLOSE SEASON PROTECTION

The species listed below do not have the full protection comparable with that of the WCA; nevertheless, they are protected during the part of the year when they breed.

Game Act 1831 Game (Scotland) Act 1772	Partridge, pheasant, black and red grouse, ptarmigan
Ground Game Act 1880 as amended	Hares and rabbits

Hares Preservation Act 1892	Hares (sale only)
Conservation of Seals Act 1970	Grey and common seals
Salmon and Freshwater Fisheries Act 1975 and Scottish Acts	Salmon, trout, other freshwater fish, eels and lamperns
Deer Act 1963 Roe Deer (Close Seasons) Act 1977	Red, sika, fallow and roe deer in England and Wales
Deer (Scotland) Act 1959 (as amended)	Red, sika, fallow and roe deer in Scotland
Wildlife and Countryside Act 1981	Schedule 2 Part I birds (q.v.)

The close season legislation also imposes restrictions on the methods by which such animals may be taken or killed at any time of year and provides for exceptions to these restrictions with or without the authority of a licence. A summary of the provisions and of the close seasons is given in Table 1; for further information see the references given in Note 1 to Table 1 and Sandys-Winsch (1984); Moss (1983) and Cooper (1984) deal with all aspects of the law relating to deer.

Outside the close season these species may be pursued, taken or killed by permitted means. A game licence is usually required in respect of hares, pheasant, partridge, grouse, black game and moor game (Game Act 1831) and woodcock, snipe, rabbits and deer (Game Licences Act 1860). Fishing is also subject to licensing under the Salmon and Freshwater Fisheries Act 1975.

CONTROL OF PESTS

The legislation relevant to the control of pests is drawn from the fields of welfare, conservation and importation together with that specifically designed to deal with vermin. The practical aspects of the application of the law in this field has been discussed in Britt (1985) and BFSS (1986), and many of the provisions are given by Sandys-Winsch (1984) and Stuttard (1986).

Specific provisions for the control of pests are provided by the Pests Act 1954, the Agriculture Act 1947, the Agriculture (Scotland) Act 1948 and the Forestry Act 1967 which provide powers to require landowners or occupiers to clear their land of rabbits, hares, deer, moles and other vermin including foxes, rats and mice.

The Animal Health Act 1981 gives MAFF powers to make provision for the destruction of wild animals for the purpose of controlling disease in animals. The Rabies (Control) Order 1974 provides for the killing of foxes in the control of a rabies outbreak and for the use of otherwise prohibited methods such as poison or trapping. Badgers have been killed or taken by MAFF under the Badgers (Control Areas) Order 1977 (which has now been repealed) on account of their alleged link with tuberculosis in cattle (Dunnet, Jones and McInerney, 1986).

Protected species, deer, badgers, birds and WCA Schedule 6 animals may be killed or taken for the protection of crops and other property or in the interests of public or animal health, although in some cases, such as WCA Schedule 1 birds, a licence may be required (see earlier).

The use of poison to kill animals is forbidden or restricted by various provisions intended to prevent suffering (see Chapter 3). In a number of situations in which legislation permits the use of poison to control pests the former are overriden. Thus, the Agriculture Act 1947 and the Prevention of Damage by Rabbits Act 1939 permit the use of gas in burrows and earths to control pests and the Grey Squirrel (Warfarin) Order 1973, made under the Agriculture (Miscellaneous Provisions) Act 1972, permits the use of warfarin to kill grey squirrels.

The use of traps and snares is restricted by the WCA and other conservation and close season legislation. For example, the WCA forbids the use of self-locking snares altogether and the use of any kind of snare to take or kill a Schedule 6 wild animal except, in either case, under licence.

The Pests Act 1954 and the Agriculture (Spring Traps) (Scotland) Act 1969 restrict the type of spring traps which may be used to those specified (as to type and species to be caught) in the Spring Traps (Approval) Orders 1975 and 1982. Spring traps used for taking rabbits or hares may only be set in rabbit holes except under licence or as part of a rabbit clearance order. The Protection of Animals Acts and the WCA require traps which have been set to be inspected at least once a day.

Breakback traps of types approved by the Small Ground Vermin Traps Order 1958 may be used to kill rats, mice and other smaller pests.

Harris (1985) points out that the killing of larger species in populated or urban areas by shooting is restricted by the limitations on the use of firearms in such places by the Highways Act 1959, Town Police Clauses Act 1847, Cemeteries Clauses Act 1847 and the Burial Act 1852.

The Destructive Imported Animals Act 1932 provides powers to control the keeping and importation of certain non-indigenous animals which have become established in Great Britain. A licence must be obtained from MAFF in order to keep mink, musk rat, musquash, coypu, rabbits (other than the European rabbit) and grey squirrels and their importation together with that of hares is also restricted (see Chapter 5). The occupier of land on which these animals are found must report their presence to MAFF. It is an offence to release these species or to allow them to escape into the wild.

The importation, keeping or release of certain species of live foreign fish or molluscs or their eggs may be prohibited or controlled by licence on environmental grounds under the Import of Live Fish (England and Wales) Act 1980 and the Import of Live Fish (Scotland) Act 1978.

TRADE CONTROL

National Trade Control

The sale of legally protected species in Great Britain must be authorised by a DOE licence under the WCA or, in the case of badgers, by the NCC (Badgers Act 1973). This may be one issued to an individual authorising one or more transactions or may be an OGL which enables any person to sell within the scope and conditions of the OGL. This avoids the need for personal application although the OGL may stipulate notification of the transaction to the DOE, record keeping or inspection.

Sale includes hire, barter and exchange and offences extend to buying and actual sale, and offering, exposing or possessing protected species, their parts or derivatives for sale. Advertising sale or purchase is subject to the same restrictions as sale itself.

The sale of protected wild animals is governed by WCA section 9(5) and the sale of wild birds by WCA section 6. The latter provides certain exceptions to the requirement for sale of live birds under licence. Those listed in Schedule 3 Part I may be sold without a licence provided that they have been captive-bred and close ringed in accordance with the Wildlife and Countryside (Ringing of Certain Birds) Regulations 1982 (DOE, 1982). The list is limited to a few British species of which there is a self-sustaining population in captivity and includes the blackbird, goldfinch, barn owl and yellowhammer.

Certain dead wildfowl and game birds listed in Schedule 3 Part III may be sold between 1 September and 28 February each year and dead feral pigeons and woodpigeons may be sold at any time (Schedule 3 Part II). Restrictions on sale are applied to game birds to which the WCA does not apply (see earlier) by the Game Acts.

Dead wild birds of other species may be sold by persons registered under the Wildlife and Countryside (Registration to Sell etc. Certain Dead Wild Birds) Regulations 1982 (DOE, 1984a). This facility is particularly designed for taxidermists, museums and those dealing in antiques and provides for the recording of sales and the marking of specimens.

In all other cases the sale of dead protected animals must be authorised by personal licence or OGL. OGLs have been issued permitting the sale within Great Britain of various live birds of prey which have been captive-bred and ringed under WCA section 7 (see earlier). An OGL authorises the sale of feathers or feathered skin from dead wild birds of Schedule 2 Part II and Schedule 3 Part III. OGLs are issued for limited periods only and are subject to variation.

The Badgers Act 1973 sections 3 and 9(2) provides that the sale of badgers must be authorised by a licence issued by the NCC or MAFF as appropriate.

The sale of game is restricted to the relevant close seasons and must be carried out by licensed game dealers. Those killing game under a game licence must sell to a licensed game dealer (Game Act 1831, Game Licences Act 1860, Deer Act 1980). Indigenous hares must not be sold between 1 March and 31 July (Hares Preservation Act 1892).

The sale of species listed in Schedule 4 Endangered Species (Import and Export) Act 1976 (ESA) (which appears in WCA Schedule 10) or any immature stage, part or derivative thereof requires a DOE licence unless it involves an item imported before the WCA was passed (30 October 1981). These restrictions refer also to other forms of trade, possession and transportation or displaying for trade of anything which has been illegally imported (DOE, 1982).

While the EEC Regulation 3626/82 (as amended) prohibits the display for trade of any A(1) specimen (see later) (as well as trade itself and keeping, offering or transporting for sale) this is subject to exceptions which may be applied in individual cases by the DOE, e.g. specimens bred in captivity or intended for research, teaching or breeding.

International Trade Control

International trade in animals, particularly wild exotic species, is subject to control of their importation and exportation. This is intended to reduce the exploitation of species which are at risk of extinction.

The Convention on International Trade in Endangered Species of Fauna and Flora (known as CITES or the Washington Convention), ratified by many countries throughout the world, aims to regulate this trade. The terms of the Convention are put into effect by the national legislation of individual countries, in the UK by the ESA (as amended by WCA Schedule 10 and elsewhere). CITES controls also take effect in the UK directly through the EEC Regulation 3626/82 (as amended) which controls the traffic in endangered species within the EEC.

The ESA and EEC Regulations prohibit the importation and exportation, unless authorised by DOE licence, of both live and dead specimens of the many species specified in Schedule 1 to the Act and certain derivatives listed in Schedule 3. See Appendix 3 (Note 21). In respect of mammals, birds, reptiles and amphibians, they are "reverse listed" or, in other words, all species are subject to the Act except those listed in the Schedule. In respect of fish, insects and molluscs only those named in the Schedule are protected.

For the purposes of administration (see DOE, 1984b) the species covered by the Act are divided into several categories, in respect of which different licensing conditions exist:

A(1)	Endangered species (CITES Appendix I, EEC Regulation Annex C, Part I) No trade dealings are allowed, but exceptions may be made for captive-bred specimens and non-commercial dealings between scientific institutions Advice of a scientific authority is obtained by the DOE before a licence is granted Live imports have to be kept at specified premises and subject to inspection Corresponding export documents are required
A(2)	Threatened species (CITES Appendix II, EEC Regulation Annex C, Part II) Commercial trading is permitted provided that it is not detrimental to conservation Evidence of country of origin and compliance with its law must be supplied Corresponding export documents are required
B1	Vulnerable species (CITES Appendix III, EEC Regulation Annex C, Part III) Trading is permitted Licences are normally available The conservation status of the species to be imported must be satisfactory Export documentation is sometimes required
Other species	HM Customs and Excise may require an importer to declare, on the arrival of a specimen, that there is no legal restriction on its importation Licences are normally issued for single consignments and remain valid for six months
Species for release to wild	A licence authorising release must be obtained under WCA before an ESA licence is issued
EEC	The EEC Regulation provides for the recognition of CITES licences issued by member countries Specimens taken from the wild within the EEC and

moved between EEC countries also require a certificate of origin

While specimens imported from outside the EEC do not normally require additional permits on movement between member countries, stricter controls may be imposed by national legislation and the UK requires licences for each import and export, even between member countries, of diurnal birds of prey, bird plumage and eggs, vicuna products

The regulations may, from time to time, restrict the importation into the EEC countries of certain species

Violation of the licensing provisions of the ESA may lead to prosecution. A person in possession of an animal being imported or exported must be able to prove that this is permitted under the Act; failure to do so makes him liable to forfeiture of the animal under the Customs and Excise Act 1952 (ESA s. 1(8)). A power of entry (as in the WCA (see earlier)) exists to ascertain whether animals kept in premises have been imported legally (ESA s. 1(10); WCA Schedule 10). Forfeited animals may be returned to the wild or kept at suitable premises at the expense of the importer or exporter or of the person having possession or control of them when they were forfeited (ESA s. 1(9); WCA Schedule 10). Similar provisions are made in respect of the EEC Regulations by the control of Trade in Endangered Species (Enforcement) Regulations 1985.

TABLE 1 Close Season Protection

Legislation	*Species*	*Protection*		*Prohibited** methods*	*Exemptions/ Defences*	*Licences*
Game Act 1831 s. 3 as amended by Game Act 1970 and Game (Scotland) Act 1772 s. 1	GAME BIRDS	CLOSE SEASONS Offence to take or kill during: Sunday Christmas Day close seasons:	NIGHT* Night Poaching Acts 1828 and 1844 Offence to take or destroy game birds and bustards	Use of: dog guns net other means to take or kill on Sunday, Christmas Day		
	Partridge	2 Feb to 31 Aug				
	Pheasant	2 Feb to 30 Sept				
	Black(heath) game	11 Dec to 19 Aug (30 Aug in places)				
	Grouse(moor) game Ptarmigan (in Scotland)	11 Dec to 11 Aug				
Game Act 1831	Bustard	1 Mar to 12 Aug				
s. 3		OTHER at any time →		Laying poison for game at any time		
s. 4	Dead game birds	Offence: sale in close season (less 10 days)			Sale of live birds for: rearing exhibition	
s. 24	Eggs of game birds also swan, wild duck, teal, wigeon	Unauthorised taking, destruction of eggs at any time				

Wildlife and Countryside Act 1981 s. 27(1)	Partridge Pheasant Black game Grouse Ptarmigan	Act does not apply to these birds except →	s. 5 applies at ANY time see pp. 126–127	As for all wild birds also: s. 5(5)(c) permits use of net or cage to take game birds for breeding purposes	s. 16 as for all wild birds
Wildlife and Countryside Act	WILD BIRDS	CLOSE SEASONS	s. 5 applies at ANY time (see pp. 126–127)	As for all wild birds also: s. 5(5)(b) permits use of nets in duck decoy in use before 1954	s. 16 as for all wild birds
s. 2(4)(a)	Capercaillie	1 Feb to 30 Sept			
	Woodcock (England & Wales)	1 Feb to 30 Sept			
	(Scotland)	1 Feb to 31 Aug			
s. 2(4)(b)	Snipe	1 Feb to 11 Aug			
s. 2(4)(c)	Wild duck Wild geese (below highwater marks)	21 Feb to 31 Aug			
s. 2(4)(d)	Other species on Sched. 2 Pt. I†	1 Feb to 31 Aug			
		DURING close season:			
s. 1, s. 1(7)	Sched. 1 Pt. II† birds	"special" protection			
s. 1	Sched. 2 Pt. I† birds	"ordinary" protection			

TABLE 1 Continued

Legislation	*Species*	*Protection*	*Prohibited** methods*	*Exemptions/ Defences*	*Licences*
s. 2(1)	Sched. 1 Pt. II† Sched. 2 Pt. I† birds	OUTSIDE close season: may be killed or taken. It is not an offence to injure in an attempt to kill			
s. 2(3)		s. 2(1) does not apply in Scotland on Sundays or Christmas Day or in some parts of England and Wales on Sundays			
s. 2(6)	Sched. 1 Pt. II† Sched. 2 Pt. I† birds	Close season protection may be provided for max. 14 days by DOE Order (e.g. in bad weather)			
		OTHER			
s. 6(6)	Sched. 3 Pt. III† birds	Sale prohibited except: dead birds 1 Sept to 28 Feb		s. 6(2) sale by person registered under s. 6(7)	s. 16 sale under licence (OGL or personal) of dead or live birds

England and Wales Deer Act 1963 s. 1(1)	DEER Acts apply to all deer, wild or enclosed or farmed††	CLOSE SEASONS	NIGHT*			
		Offence to take or wilfully to kill	s. 2 offence to take or wilfully to kill ANY species of deer at ANY time of year	s. 3(1)(a) setting trap, snare, poison, stupefying bait likely to injure	s. 10(1) defence to s. 1, s. 2, s. 3(1)(a)(b) (NOT (c)) for any act to prevent suffering of injured or diseased deer	NCC licence: s.11 to use nets, traps, drugs, to take deer alive for: translocation or scientific purposes
	Red stags Sika stags Fallow bucks	1 May to 31 July				
Roe Deer (Close Seasons) Act 1977	Roe bucks	1 Nov to 31 Mar		s. 3(1)(b) using net to take, kill	s. 10(2) exemption from s. 1, s. 2 for control of deer authorised by MAFF	
	Red hinds Sika hinds Fallow doe Roe doe	1 Mar to 31 Oct		s. 3(1)(c) to take, kill, injure, using: (i) air weapon shotgun rifle of prohibited calibre (ii) arrow spear (iii) missile carrying:	s. 10(3)(4) specified calibre of shotgun permitted for euthanasia or as slaughtering instrument s. 10A(1)–(3) (WCA Sched. 7) defence for close season shooting in or with permitted	

TABLE 1 Continued

Legislation	*Species*	*Protection*		*Prohibited*** *methods*	*Exemptions/ Defences*	*Licences*
Deer Act 1987	Any species	s. 10(2A) close seasons not applicable to farmed (enclosed and marked) deer		poison stupefying drug WCA s. 11(1)(b) crossbow s. 3(2) vehicle to drive or shoot deer	shotgun by authorised person to protect crops etc.	
		OTHER Offence to:				
Deer Act 1980 s. 2(1)(a)		sell venison during last 10 days of close season			sale by licensed game dealer	
s. 2(1)(b)		sell venison other than to licensed game dealer				
s. 2(2)		sell deer subject of offence under 1963, 1980 Acts				
Scotland Deer (Scotland) Act 1959 s. 21 as amended by: Deer (Amendment) (Scotland) Acts 1967 and 1982		CLOSE SEASONS	NIGHT*			
	Red stags Sika stags Hybrid stags	Offence to take, wilfully kill or injure: 21 Oct to 30 June	s. 23(1) offence to take, wilfully kill or injure ANY deer at ANY time of	s. 23(2)–(5) offence to take, wilfully kill or injure ANY deer exceptions: shooting	s. 33 defence for: anything done to prevent suffering in a sick or injured deer or	s. 33(3B) killing or taking in close seasons for scientific purposes (with written authority

	Roe bucks	21 Oct to 31 Mar	year	(only permitted guns)	orphaned calf, fawn, kid	of Scottish Office)
	Fallow bucks	1 May to 31 July		taking alive must not cause unnecessary suffering, taker must have right to take the deer	exceptions for: close season or night shooting on specified land to protect crops etc. close season, night shooting of marauding deer	Agriculture Act 1948 s. 39 control of deer authorised by DAFS
Deer (Close Seasons) (Scotland) Order 1984 and Deer (Firearms, etc.) (Scotland) Order 1985 (see Anon, 1982) for amended version of 1959 Act)	Red hinds, Sika hinds, Hybrid hinds	16 Feb to 20 Oct				
	Fallow doe, Roe doe	1 April to 20 Oct				
				Vehicles, aircraft not to be used to drive or shoot deer on enclosed land	s. 21(5A) close seasons not applicable to farmed (enclosed and marked) deer	
				Carriage of deer inside aircraft permitted; outside only under veterinary supervision		
s. 25D	Any species	OTHER Venison must be sold by a licensed venison dealer				
Conservation of Seals Act 1970	SEALS	CLOSE SEASONS Offence wilfully to take, kill or injure:		s. 1(1) offence: to take, kill or	s. 9 defences to s. 2, s. 3:	s. 10 licence available to: take, kill seals

TABLE 1 Continued

Legislation	*Species*	*Protection*	*Prohibited** methods*	*Exemptions/ Defences*	*Licences*
s. 2(1)	Grey seal Common seal	1 Sept to 31 Dec 1 June to 31 Aug	injure ANY seal by any firearm except rifle of permitted calibre	(a) taking disabled seal for care and subsequent release (b) unavoidable killing, injury arising from lawful action (c) preventing damage to nets, fish defence to s. 1–s. 3 euthanasia s. 17(2) exemption for any act outside GB territorial waters	for scientific, educational purposes, zoological collections, preventing damage to fisheries, population control, protecting flora, fauna
s. 3	Both species	at any time in a designated area	to kill, take ANY seal by poison		
Scotland*** England & Wales Salmon and Freshwater Fisheries Act 1975	FISH	CLOSE SEASONS	s. 21	s. 19	s. 19
	Freshwater fish or eels	Set by local river authority bye-laws*** if none Act provides: illegal to take or kill in any inland water 14 Mar to 16 June	use of eel traps etc. in water used by salmon or migratory trout: 31 Dec to 25 June Lampern traps permitted on weirs: 1 Aug to 1 Mar	rights of several fishery	written authority of MAFF or river authority: for scientific purposes

Sched. 1	Rainbow trout	Bye-law*** only		s. 19 rights of several fishery	As above
ss. 1–8	Freshwater fish		ss. 1–8 at ANY time: firearms certain lines, nets, instruments fish roe explosives, poison electric fishing		As above
s. 2	Freshwater fish		illegal to take, kill immature or unclean fish		As above
Sched. 1	Salmon	Set by bye-laws*** if none: nets: 31 Aug to 1 Feb rod and line: 31 Oct to 1 Feb basket traps: 31 Aug to 1 May Each week: 6 am Sat to 6 am Mon Sale of fresh salmon: 31 Aug to 1 Feb	s. 20 any obstruction to the passage of salmon ss. 1–8 as above	s. 22 caught outside UK; caught inside UK as mature clean fish other than by rod and line; processed within close season or abroad	s. 19 written authority of MAFF or river authority: artificial propagation, scientific purposes

TABLE 1 Continued

Legislation	*Species*	*Protection*	*Prohibited** methods*	*Exemptions/ Defences*	*Licences*
Sched. 1	Trout (other than rainbow trout)	Set by bye-laws***, If none: nets: 31 Aug to 1 Mar rod and line: 30 Sept to 1 Mar Each week: 6 am Sat to 6 am Mon Sale: 31 Aug to 1 Mar	s. 20 any obstruction to the passage of migratory trout ss. 1–8 as above	As above	As above also for stocking waters
Game Act 1831 as amended by Game Act 1970	GROUND GAME Hares	Offence to take or kill on: Sunday Christmas Day	s. 3 use of dog, gun, net or other means to take or kill on Sunday, Christmas Day laying poison for game at any time		
Hares Preservation Act 1892	Hares	Offence to sell: 1 Mar to 31 July		s. 3 imported hares	

Night Poaching Acts 1828 & 1844	Hares Rabbits		Offence: unlawfully take or destroy	
Ground Game Act 1880 s. 6	Hares Rabbits	NIGHT* shooting	Spring traps unless set in rabbit holes poison	night shooting by occupier or one other person duly authorised
Agriculture (Scotland) Act 1948 ss. 50, 51 as amended by WCA s. 12 Sched. 7				

This table is necessarily only a guide to the law. For further detail reference should be made to BASC (1983, 1984), Cooper (1984), Gregory (1974), Parkes (1983), the sources of law mentioned in Chapter 1 and the relevant Acts of Parliament.

The Acts cited apply to England, Wales and Scotland unless otherwise stated.

The taking or killing of most animals subject to close seasons requires a game or fishing licence and is subject to the law relating to trespass and poaching.

*Night: between one hour after sunset and one hour before sunrise.

**Prohibited methods: such methods of killing or taking are prohibited at all times (not only during the close seasons) unless otherwise stated.

***Fish:

(a) The Scottish law differs to some extent. It is regulated by several Acts from the Salmon Act 1691 to the Salmon Fisheries (Scotland) Act 1976. An updated version of this legislation is provided in *Statutes in Force* (HMSO, continuing).

(b) For details of river authorities, their licensing requirements and the close seasons set by bye-laws see Orton (1984).

† See Analysis of Birds Protected by the Wildlife and Countryside Act 1981, p. 148.

†† Some species of deer are present but do not have close seasons (e.g. muntjac).

†ANALYSIS OF BIRDS PROTECTED by the Wildlife and Countryside Act 1981

Species	*Protection* (XX "special" Sched. 1 Pt. II, X "ordinary" Sched. 2 Pt. I)	*Close seasons* s. 2(4)(a)–(c) various dates	s. 2(4)(d) 1 Feb to 31 Aug	*Sale* 1 Sept to 28 Feb Sched. 3 Pt. III
Capercaillie	X	X		X
Coot	X	also	X	X
Duck, tufted	X	wild duck	X	X
Gadwall	X	wild geese	X	
Goldeneye	XX	below	X	
Goose, Canada	X	high	X	
Goose, greylag	X	water	X	
Goose, greylag (in parts of Scotland)	XX	mark	X	
Goose, pink-footed	X	|	X	
Goose, white-fronted (in some places)	X	|	X	
Mallard	X	|	X	X
Moorhen	X	|	X	
Pintail	XX	|	X	X
Plover, golden	X	|	X	X
Pochard	X	|	X	X
Shoveler	X	|	X	X
Snipe, common	X	X		X
Teal	X	|	X	X
Wigeon	X	|	X	X
Woodcock	X	X England and Wales	X Scotland	X

CASE

Kirkland *v*. Robinson (1986). *The Times*, 4 December 1986.

REFERENCES

Anon. (1982). Deer (Scotland) Act 1959. *Statutes in Force, Animals 3*. HMSO, London.

BASC (1983). *Know Your Law England and Wales*. British Association for Shooting and Conservation, Rossett, Clwyd.

BASC (1984). *Know Your Law Scotland*. British Association for Shooting and Conservation, Rossett, Clwyd.

BFSS (1986). *Predatory Mammals in Britain*, 4th edn (Stuttard, R.M., ed.). British Field Sports Society, London.

Beebee, T. (1983). Wildlife, the law . . . and you. *British Herpetological Society Bulletin* **1**, 55–56.

Britt, D.P. (ed.) (1985). *Humane Control of Land Mammals and Birds*. Universities Federation for Animal Welfare, Potters Bar.

Cooper, M.E. (1984). The law. In *Guidelines for the Safe and Humane Handling of Live Deer in Great Britain*. Deer Liaison Committee and Nature Conservancy Council, Peterborough.

Cooper, M.E. (1986). British legislation relating to the conservation of reptiles and amphibians. *ASRA Journal* **3**(1), 31–38.

DOE (1982). *Sale of Schedule 4 or 5 Species*. Department of the Environment, Bristol.

DOE (1983). *A Guide to the Registration of Species Listed on Schedule 4 for Keepers and Owners*. Department of the Environment, Bristol.

DOE (1984a). *A Guide to the Registration of Sellers of Dead Birds*. Department of the Environment, Bristol. Now (1987) with Supplementary Notices 1–6.

DOE (1984b). Notice to importers and exporters. *Controls on the Import and Export of Endangered and Vulnerable Species*, two supplements. Department of the Environment, Bristol.

DOE (1987). *DOE Close-ringing of Captive-bred Schedule 4 Birds. A Guide for Keepers*. Department of the Environment, Bristol.

Dunnet, G.M., Jones, D.M. and McInerney, J.P. (1986). *Badgers and Bovine Tuberculosis*. HMSO, London.

Gregory, M. (1974). *Angling and the Law*, 2nd edn. Charles Knight, London.

Harris, S. (1985). Humane control of foxes. In *Humane Control of Land Mammals and Birds* (Britt, D.P., ed.). Universities Federation for Animal Welfare, Potters Bar.

Harris, S. (1986). Badgers in law. *BBC Wildlife*, May 1986, 232.

HMSO (continuing). Fisheries. In *Statutes in Force*, Group 52, Sub-group 2. HMSO, London.

JCCBI (undated). *A Code for Insect Collecting*. Joint Committee for the Conservation of British Insects, London.

JCCBI (1986). Insect re-establishment. A code of conservation practice. *Antenna* **10**(1), 13–18.

Langton, T. (1986). *Protecting Wild Reptiles and Amphibians in Britain*. Fauna and Flora Preservation Society, London.

Low, R. (1985). Law and liberty. *Cage and Aviary Birds*. 14 September 1984, 2.
Marren, P. (1986). The lethal harvest of crayfish plague. *New Scientist*, 30 January 1986, **109**, 46–50.
Moss, R. (1983). Legislation on the handling of deer. *Publication of the Veterinary Deer Society* **1**(4), 9–19.
NCA (undated). *A Fancier's Rights in Law*. National Council for Aviculture, London.
NCC (1979). *Wildlife Introductions to Great Britain*. Nature Conservancy Council, Peterborough.
Orton, D.A. (1984). *Where to Fish 1984–1985*, 79th edn. Thomas Harmsworth, London.
Parkes, C. (1983). *Law of the Countryside*. Association of Countryside Rangers, Middleton, Saxmundham, Suffolk.
Sandys-Winsch, G. (1984). *Animal Law*. Shaw, London.
Stuttard, R.M. (1986). *Predatory Mammals in Britain*. British Field Sports Society, London.
UFAW (1985). *The Humane Control of Land Mammals and Birds*. Universities Federation for Animal Welfare, Potters Bar.

RECOMMENDED READING

See Chapter 1 for literature generally applicable.

Cage and Aviary Birds. Weekly newspaper for aviculturists.
Campbell, B. and Lack, E. (1985). *A Dictionary of Birds*. T & A.D. Poyser, Calton. (For assistance in identifying species.)
Cooper, M.E. and Cooper, J.E. (1987). Wildlife and non-domesticated species. In *Legislation Affecting the Veterinary Profession in the United Kingdom*. Royal College of Veterinary Surgeons, London.
Cooper, M.E. (1987). The law relating to the captive breeding of birds of prey. In *Breeding and Management in Birds of Prey* (Hill, D.J., ed.). University of Bristol, Bristol.
Cooper, M.E. (1987). The British law relating to raptors. *Bulletin of the World Working Group of Birds of Prey* **3**, 145–148.
Davis, R.P. (1979). *The Protection of Wild Birds*. Barry Rose, Chichester.
Denyer-Green, B. (1983). *Wildlife and Countryside Act 1981. The Practitioner's Companion*. Surveyors Publications, London.
DOE (1982). *A Guide to some Aspects of the Law on the Taking, Keeping, Sale, Exhibition and Licensing of British Birds*. Department of the Environment, Bristol.
Fox, C. (1971). *The Countryside and the Law*. David & Charles, Newton Abbot.
Heap, J. (1981–1982). *An Introduction to the Wildlife and Countryside Act 1981. Ardea, 39–44. (Journal of the Beds and Hunts Naturalists' Trust).*
NCC (1982). *Wildlife, the Law and You*. Nature Conservancy Council, Shrewsbury, now Peterborough.
NCC (1985). *Bats in Roofs*. Nature Conservancy Council, Peterborough.
RSPB (1986). *Wild Birds and the Law*. Royal Society for the Protection of Birds, Sandy.

USEFUL LEAFLETS

CoEnCo (1982). *Wildlife and the Law Number 2: Reptiles and Amphibians*. Council for Environmental Conservation and British Herpetological Society, London.
CoEnCo (in preparation). *Wildlife and the Law Number 3: Mammals*. Council for Environmental Conservation and Mammal Society, London.
DOE (1984). *Endangered Species*. Department of the Environment, Bristol.
NCC (1984). *Focus on Bats: Their Conservation and the Law*. Nature Conservancy Council, Interpretive Branch, Shrewsbury, now Peterborough.
RSPB (undated). *Information about Birds and the Law*. Royal Society for the Protection of Birds, Sandy.

8 *Health and Safety*

Salus populi suprema rex.
The greatest law is the people's safety.

(Adapted from Justinian's "Twelve Tables of Roman Law")

Since the advent of the Health and Safety at Work etc. Act 1974 (HSWA) health and safety have become a vital factor in all aspects of the work place. An understanding of the Act's implications is essential to employers, employees and others. It gives rise to regulations, guidelines and codes made under the Act itself and also requires employers to produce policy statements and "local" codes of practice or rules appropriate to particular premises. In addition, there has been a trend for national organisations to produce codes or guidelines for consideration or adoption by their constituent members.

Occupational health and safety has become a discipline in its own right and a large employer will normally appoint full-time safety officers. In addition, the Safety Representatives and Safety Committees Regulations 1977 require an employer to set up a safety committee if requested to do so by a recognised trade union and to consult safety representatives appointed by the union.

HEALTH AND SAFETY AT WORK ETC. ACT 1974

The HSWA applies to all persons at work, which includes employers, employees, the self-employed, any persons on the work premises who are not employees and any members of the general public who may be affected by activities involved in work. The category of non-employees extends to those such as tradesmen or social visitors, contractors' employees, research workers supported by an outside grant, students, visiting staff, guest speakers or voluntary workers.

The purpose of the Act is to secure the health, safety and welfare of persons at work and to protect others from risks to health and safety resulting from the activities of those at work (s.1). It is also the purpose of the Act to provide controls over dangerous materials and over the emission of obnoxious or offensive substances.

The Act took under its wing the areas such as railways, shops and factories in which there were already comparable provisions although much of the relevant legislation remains in force. However, it applied for the first time to many other work places, including those involved in education, research and veterinary practice.

General Duties

The Act imposes general duties upon an employer, employee and self-employed person which can be summarised as follows:

1. *Employer*
   (a) By section 2 of the Act an employer has a general duty to ensure so far as reasonably practicable the safety and welfare at work of all his employees
   (b) In particular this duty relates to:
      (i) The provision and maintenance of plant and machinery
      (ii) The arrangements for use, handling, storage and transport of articles and substances
      (iii) The provision of information, instruction, training and supervision in order to satisfy the general duty
      (iv) The condition of work premises and access and egress therefrom
      (v) Maintenance of the work environment
      (vi) Provision of facilities and arrangements for welfare
   (c) Every employer must provide a written statement of his general policy for health and welfare and of the means by which it is implemented and brought to the notice of employees
   (d) The employer must make provision for cooperation with employees over health and safety measures, including compliance with the Safety Representatives and Safety Committees Regulations 1977
   (e) Section 3(1) imposes a duty upon an employer to conduct his business so that non-employees are not exposed to risks to their health or safety and to provide such people with safety information

2. *Self-employed person*
   (a) A self-employed person who is also an employer has the foregoing duties in 1 above.
   (b) He must conduct his undertaking to ensure, as far as is reasonably

practicable, that he himself and persons other than his employees are not exposed to risks to their health and safety (s. 3(2)).

3. *Employee*
   Section 7 of the Act imposes a duty on an employee:
   (a) to take reasonable care for the health and safety of himself and of any other person who may be affected by his acts or omissions at work
   (b) to cooperate with his employer or other persons so far as is necessary to enable them to perform any duty imposed upon them by the Act

4. *Non-employee*
   (a) The Act imposes no specific duty upon such a person. However, it is possible to provide, as in the Health and Safety (Genetic Manipulations) Regulations 1978, that he be treated as if he were self-employed, thereby making him responsible as in 2 above
   (b) Duties *towards* non-employees:
      (i) For an employer's duties see 1(e) above
      (ii) For a self-employed person's duties see 2 earlier
      (iii) An employee has a duty to take reasonable care for the health and safety of a non-employed person (see 3(a) above)

5. *Any person*
   Section 8 provides that no person may intentionally or recklessly interfere with or misuse anything provided under the Act in the interests of health, safety or welfare

6. *The person in control of premises*
   By sections 5 and 6 of the Act a duty is imposed upon a person in control of premises in respect of:
   (a) The prevention of emission of noxious or offensive substances
   (b) The production and supply of articles or substances for use at work (this includes the provision of information to users and the testing of products). This may require experiments under the Animals (Scientific Procedures) Act 1986 (see Chapter 4))
   (c) The installation of articles for use at work

It is important to note that all duties under the Act are imposed "so far as is reasonably practicable". Thus, it is not an absolute duty to prevent any danger to health and safety but a matter of balancing the measure of the risk with other factors, such as feasibility and cost. As Redgrave says, "reasonably practicable" implies that "a computation must be made in which the *quantum* of risk is placed in one scale and the sacrifice involved in the measures necessary for averting the risk (whether in money, time or trouble) is placed in the other . . ." (Fife and Machin, 1982).

Implementation of the Employer's Policy Statement

The employer's policy statement is usually written in brief and general terms with reference to detailed guidelines and codes which relate to specific kinds of work, parts of premises (e.g. an animal house, a laboratory or an operating theatre), a type of work (say, field studies) or particular substances. There may also be codes or instructions relating to matters common to all parts of an employer's premises, such as fire safety precautions or first aid procedures. Such literature must be readily understood and provision made for those who have limited comprehension of the language or the risks involved.

REGULATIONS, CODES OF PRACTICE AND GUIDELINES

Regulations together with, in some cases, codes of practice and guidance notes have been produced on the following subjects which may be applicable to work with animals:

- Health and Safety (Genetic Manipulations) Regulations 1978
- Health and Safety (First Aid) Regulations 1981
- Health and Safety (Dangerous Pathogens) Regulations 1981
- Safety Signs Regulations 1980
- Classification, Packaging and Labelling of Dangerous Substances Regulations 1984
- Notification of New Substances Regulations 1982
- Notification of Installations Handling Hazardous Substances Regulations 1984
- Ionising Radiations Regulations 1985

Approved codes of practice have also been published in respect of ionising radiation (HSC, 1985a) first aid (HSE, 1982) and work at zoos (HSC, 1985b). Responsibility under the latter has been transferred from the Health and Safety Executive to local authorities because of their responsibility for the inspection of zoos under the Zoo Licensing Act 1981 (see Chapter 3).

The new radiation provisions came into force on 1 January 1986 and existing non-statutory codes of practice are replaced or revised by Regulations and Approved Codes of Practice. Those of particular relevance to this book which were in use before that date are the following.

Guidance Notes for the Protection of Persons exposed to Ionising Radiation in Research and Teaching (HSE, 1968)

Radiation Safety in Veterinary Practice (MAFF, 1974)
Code of Practice for the Protection of Persons against Ionising Radiation arising from Medical and Dental Use (DHSS, 1972)

Codes of practice and guidance notes have been produced widely throughout all fields of employment. Unless these have been approved by the Health and Safety Commission, which is rare, they have no legal standing. Such documents may be "local", i.e. produced for a particular employer's premises, or they may be compiled by an organisation for voluntary application by its members or others. Relevant publications include MAFF (1978), DHSS (1980), BVA (1981), Nichols (1983) and Collins (1985). An example of a code of practice for an animal house is given by Seamer and Wood (1981).

Statutory regulations have legal status and the breach of them is an offence. Codes of practice produced or approved by the Health and Safety Commission have a special standing in that failure to observe them is not an offence as such; however, they may be produced in evidence when establishing a contravention of some statutory provision such as the HSWA itself or a regulation made under it. Failure to comply with a code relevant to an alleged contravention is accepted as proof of the contravention unless the accused can show that he complied with the point of law in question in some other way (HSWA s. 17).

Unless the regulations specifically so state, breach of regulations may also form the basis of a claim in civil law, e.g. in negligence (see later). While codes of practice, guidance notes or guidelines do not have any legal force, production of, and adherence to, them are indicative of an intention to comply with the duties imposed by the HSWA and by the civil law.

ENFORCEMENT OF HEALTH AND SAFETY LEGISLATION

The enforcement of health and safety legislation is the responsibility of the Health and Safety Executive through its Health and Safety Inspectorate. Inspectors visit premises to ensure compliance with the legislation and also to give advice. In the case of a breach they have power to issue an improvement notice requiring the remedy of the fault. In cases of severe risk to health or safety a prohibition notice may be served requiring the cessation of some aspect of work until the improvements required by the notice have been made. Prosecutions may also be brought for activities which do not comply with the Act, regulations or a notice. There is also power to seize (and render harmless) a substance or article which an inspector considers to be causing imminent danger or serious personal injury.

OTHER SAFETY PROVISIONS

Dangerous Substances

There are numerous provisions regarding dangerous substances, some of which have already been mentioned because they are covered by regulations made under the HSWA. Others which may be relevant to work with animals include the following.

Hazardous substances	The proposed Control of Substances Hazardous to Health Regulations will apply to most chemical substances and will extend to people working with animals
Carcinogens	The Carcinogenic Substances Regulations 1967 require that a licence is obtained to use certain material for scientific research. While these Regulations have only a limited effect, the use of carcinogens may be more strictly controlled under the Regulations above
Medicines	The restrictions on the supply of controlled drugs and other prescription only medicines are discussed in Chapter 6
Poisons	Poisonous substances which are not medicines are controlled by the Poisons Act 1972 and the Poisons Rules 1982. Poisons must be sold through pharmacies except for Part II poisons which may be supplied by listed sellers to a person or institute concerned with scientific education or research for such purposes (Pearce, 1984)
Radioactive substances	The Radioactive Substances Act 1948 and 1960 and numerous regulations and codes, in particular the Ionising Radiations Regulations 1985, deal with the handling, storage, use, disposal and transportation of radioactive material

Fire Prevention

The Fire Precautions Act 1971 applies to certain places of work, particularly factories (these are very widely defined) but, although not yet extended to academic and comparable institutions, this could be so in the future. The Fire Certificates (Special Premises) Regulations 1976 apply to places where there are special hazards, such as highly inflammable substances or radioactive material.

The Act requires that there should be a fire certificate in force in respect of such premises. The fire authority must have inspected the premises and been satisfied that fire escapes, fire-fighting equipment and warning procedures are reasonable in the circumstances. The certificate must give details of these matters and must be kept on the premises to which it refers. A certificate may impose conditions as to the provision of a means of escape, of training and instruction of employees and other fire precautions, including the restriction of numbers in a building.

Waste Disposal

The Control of Pollution Act 1974 provides that controlled waste — household, industrial or commercial waste — must be disposed of in accordance with local authority arrangements. Commercial waste includes "waste from premises used wholly or mainly for the purposes of a trade or business . . .".

The Control of Pollution Act (s. 12) imposes upon district councils the duty to collect household waste without charge (unless a charge is permitted by the 1987 Regulations (see below)) and a duty to collect commercial waste when requested to do so by the occupier (a charge may be made). Councils have a discretion whether to collect industrial waste from factories (as defined by the Factories Act 1961) (Webster, 1981).

The Control of Pollution (Collection and Disposal of Waste) Regulations 1987 specify in detail the kinds of waste which fall within the three main categories.

It is an offence to deposit waste so that it causes an environmental hazard, i.e. if it is likely to become a material risk to human or animal health or to threaten water supplies. It is an offence to deposit waste unless authorised by a disposal licence issued by a local authority.

The Control of Pollution (Special Waste) Regulations 1980 provide for written notification to be given to waste disposal authorities of the disposal of "special waste". This includes prescription only medicines and substances with specified risks.

The Act also forbids the drainage of trade effluent into the public sewer except with the consent of the relevant water authority; the Public Health (Drainage of Trade Premises) Act 1937 (amended by the Public Health Act 1961) requires the notification of discharge of effluent from premises used for trade, industry, science and research. Notification must give the water authority details of the material to be discharged.

Postage of Pathological Material

Postage of pathological material must comply with the Post Office's leaflets K681 (inland mail) or DSO 61 (overseas). The former sets out the requirements for packing such material, including the use of leak-proof containers and absorbent material; the package must be marked "fragile with care" and "pathological specimen" and sent by first class mail. The instructions for international postage are complex and reference should be made to the leaflet or to the *Post Office Guide* (Post Office, annual).

ACCIDENTS AT WORK

Notifiable Accidents

The Social Security Act 1975 provides for compensation in the form of industrial injury benefit, industrial disablement benefit and industrial death benefit to be paid to those suffering death or injury arising out of or in the course of employment. The Act requires an employee to report such an accident to his employer who must investigate the accident and make a report to the Department of Health and Social Security (DHSS). An employer must keep an accident book in the format approved by the DHSS in which employees may record accidents involving personal injury (Social Security (Claims and Payments) Regulations 1975).

Serious accidents including "dangerous occurrences" which do not cause injury and cases of illness must be reported to the HSE under the Reporting of Injuries, Diseases and Dangerous Occurrences Regulations 1985.

Health and Safety at Work etc. Act 1974

The circumstances surrounding a personal accident at work may constitute an offence under the HSWA or regulations made under it. This may lead to prosecution of the person responsible and, on conviction, to a fine or imprisonment.

Civil Law

The same accident may also give rise to a claim in civil law by the injured person. A claim may be made on the grounds mentioned below. To succeed the facts and liability must be proved. The defendant may seek to show that the plaintiff consented or contributed to the accident, thereby reducing the former's liability and the damages which might be awarded.

Vicarious liability
An employer is responsible for the negligence of his employee which occurs in the course of his employment.

Employers' liability
At common law an employer is required to take reasonable care for the safety at work of his employees by providing competent staff, safe premises and a safe system of work. Since the Law Reform (Personal Injuries) Act 1948 an employer has been liable for injuries to employees caused by the negligence of a fellow employee. The Employers' Liability (Defective Equipment) Act 1969 renders an employer liable for injury caused to his employee by defective equipment provided for the purpose of his business even though the fault is the responsibility of a third party such as the manufacturer.

Breach of statutory duty
A person (employer or other) who fails to comply with statutory requirements may also be liable in civil law for injury or damage which are the consequence of that failure. Regulations made under the HSWA give rise to such liability (unless it is expressly excluded) but the Act itself cannot be the basis of a claim under this head.

Negligence
Liability arises in negligence when it is reasonably foreseeable that some harm will be caused and there is a failure to prevent it which causes damage, injury or death. In the context of accidents at the work place, such liability may arise either out of, or independently of, circumstances which are also a breach of the HSWA.

Other forms of liability
There is a duty upon the occupier of premises to ensure that a person lawfully on those premises does not suffer any harm caused by the state of the land or buildings. Under the Occupiers' Liability Act 1984 the occupier who knows that trespassers come on his property must take reasonable steps to warn them of hazards or to discourage them from taking risks of personal injury

(Haley, 1984). The duty in respect of trespassers is less than that towards lawful visitors. In either case the degree of protection required will vary from simple notices to, for example, strong fencing or reinforcement and will depend on individual circumstances. Under the 1957 Act the occupier must take extra care for the safety of children, even if they are trespassers.

The Animals Act 1971 makes the keeper of a non-domesticated species liable for any harm it causes. In the case of domesticated animals the keeper is liable for damage which is due to some unusual propensity of the animal of which the keeper is already aware (see Chapter 2).

Protection from Liability

Bringing or defending a claim in civil law is costly and time consuming. It is therefore normal to ensure that potential liabilities are covered by insurance. However, under pressure from the cost of insurance, premiums and a reluctance on the part of insurers to provide cover, it has become common to devise methods of transferring or avoiding the burden of liability or its cost.

While these matters are of general application, some aspects have particularly concerned those who join or lead expeditions, field study groups or site visits and those who grant such people access to land or premises (see Berridge (1986), Gray (1986) and Cooper (1987)).

Insurance

Since it is impossible ever to be certain that all legal obligations will be satisfied all the time, be it under the safety legislation or in civil law, or that accidents will not occur, it is essential to have the protection of insurance. This applies to both the person (often an employer) at risk of being liable for an accident or the person who may suffer injury.

Crown authorities such as government departments or ministers and public (i.e. local authorities and government) agencies do not normally take out insurance since they have such extensive liabilities. They act as their own insurer.

An employer must insure against liability towards his employees in respect of injury or disease arising out of, or in the course of, their employment (Employers' Liability (Compulsory Insurance) Act 1964) although in fact he is likely to insure against a much wider range of risks.

Whilst one cannot insure against criminal prosecution because it is contrary to public policy, it is possible to effect cover for the cost of defending a case and the consequences of conviction.

When insurance cover is sought or reviewed all aspects of civil liability should be considered in conjunction with the need for personal and property insurance. It is important to consider the sum insured, the risks involved, the persons covered and whether legal costs are included in the policy. The last can be greater than damages recovered or may represent the expense of defending a case.

The type of insurance protection held may be critical to the amount recoverable in the event of an accident. With liability insurance the amount will depend on the extent of liability, assessed by a court or settled between the parties, subject to the overall limit on the policy, the balance remaining the responsibility of the insured. A personal injuries policy will give rise to payment of a specified amount for named injuries (e.g. loss of a limb) or death without question of liability. This amount would normally be much lower than that which could be claimed against a person who caused the injury or death. However, an effective claim depends on establishment of liability whereas a personal injuries claim is not dependent on proving fault.

The question of insurance has raised a number of problems in connection with educational bodies where insurance may be effected by the administrative side of an institution. It is important that those who require insurance cover, particularly teachers and pupils or lecturers and students, ensure that the policy actually meets their needs, particularly when they are working outside the premises of the institution, and that they communicate any special risks or disabilities which should be disclosed to the insurer. The distinction between a disclaimer and indemnity, which is discussed below, must be understood when assessing the incidence of liability. If a disclaimer is signed liability is negated and no action can be taken for compensation for injury or death unless the Unfair Contract Terms Act restricts the effect of the disclaimer; by contrast, an indemnity does not prevent a claim for damages being made and won or settled out of court but passes the ultimate responsibility for payment to the person who gave the indemnity.

Disclaimer

Express exclusion of liability may be considered as a means to avoid liability. This, however, is substantially restricted by the Unfair Contract Terms Act 1977 (ss. 1 and 2) so far as liabilities arising in the course of business or in connection with the occupation of premises for business purposes are concerned. Any attempt by contract or by general notice to exclude liability for causing death or personal injury by breach of a common law duty to take reasonable care or exercise reasonable skill (e.g. in negligence, employer's liability or vicarious liability) or under the Occupiers' Liability Act 1957 (in respect of dangerous premises) is ineffective. In the case of any other form of loss or damage, liability can only be excluded so far as is considered reasonable by a court (Rogers and Clarke, 1978). Thus, an occupier may put a notice on his premises excluding liability for injury to persons and property but it is ineffective as to the former and only valid as to the latter so far as is reasonable.

While in such situations it is difficult to exclude liability, the Occupiers' Liability Act 1984 has provided that the granting of access to business premises (e.g. those used for commercial, industrial or agricultural purposes) does not count as a business liability when it is provided for recreational or educational purposes. Thus, the occupier of, for example, a factory or commercial quarry is now able to exclude liability for loss or damage suffered, as a result of the dangerous state of his premises, by the members of a college field study group when they visit his premises. The only exception to this is that the occupier whose actual business is the granting of access for educational or recreational purposes, e.g. the proprietor of a sports or field studies centre, cannot exclude liability towards those using his facilities.

Indemnity

With the aim of mitigating the burden of liability, an indemnity against claims may be required as a condition of entry to, or use of, premises or the provision of goods, facilities or services.

Again, under the Unfair Contract Terms Act (s.4) there is some limitation on this. It provides that a person dealing as a consumer cannot by contract be required to indemnify another person who is acting in the course of business for liability arising in a common law duty of care or under the Occupiers' Liability Act unless it is reasonable in the circumstances (Rogers and Clarke, 1978). Such clauses are common in the provision of access to premises for field trips and the effect of this is that a private individual (e.g. a student or a private club) cannot be held to an indemnity by a company which has asked for one as part of a contract granting access to premises for a field or site visit, whereas a company or, probably, an academic institution (if it is considered a business and there is some doubt about this (Rogers and Clarke, 1978)) would not have the benefit of the Act. However, someone who is not acting in a business capacity, for example, when allowing a person or group to study in his private house or garden can insist on an indemnity.

Caution

All the foregoing areas of contract, liability and insurance are complex and professional advice should be sought according to the needs of individual circumstances.

REFERENCES

BVA (1981). *Health and Safety at Work Act. A Guide for Veterinary Practices.* British Veterinary Association Publications, London.

Berridge, J. (1986). Insurance for expeditions. In *Expedition Planners' Handbook and Directory 1986/7.* Expedition Advisory Centre, Royal Geographical Society, London.

Collins, C.H. (ed.) (1985). *The Law and Biological Laboratories.* Wiley, Chichester.

Cooper, M.E. (1987). An outline of legal requirements for fieldwork. In Symposium on the Safety of Fieldwork Activities. *Safety Digest, 19.* Universities Safety Association, Manchester.

DHSS (1972). *Code of Practice for the Protection of Persons against Ionising Radiation arising from Medical and Dental Use.* HMSO, London.

DHSS (1980). *Code of Practice for the Prevention of Infection in Clinical Laboratories and Post-mortem Rooms.* HMSO, London.

Fife, I. and Machin, E.A. (1982). *Redgrave's Health and Safety in Factories*, 2nd edn, and supplement. Butterworth; Shaw, London.

Foster, S.J. (1986). The Ionising Radiations Regulations 1985: what they mean in practice. *Veterinary Record* **117**, 489–490.

Gray, R. (1986). Legal considerations. In *Expedition Planners' Handbook and Directory 1986/7.* Expedition Advisory Centre, Royal Geographical Society, London.

Haley, M.A. (1984). The uninvited entrant and the Occupiers' Liability Act 1984. *Law Society's Gazette* **81**, 1594–1595.

HSC (1985a). *The Protection of Persons Against Ionising Radiation arising from any Work Activity. Approved Code of Practice.* HMSO, London.

HSC (1985b). *Zoos — Safety, Health and Welfare Standards for Employers and Persons at Work. Approved Code of Practice and Guidance Note.* HMSO, London.

HSE (1968). *Guidance Notes for the Protection of Persons exposed to Ionising Radiation in Research and Teaching.* HMSO, London.

HSE (1982). *First Aid at Work.* HMSO, London.

MAFF (1974). *Radiation Safety in Veterinary Practice.* Ministry of Agriculture, Fisheries and Food, Pinner.

MAFF (1978). *Safety Precautions for the Use in Veterinary Laboratories of the Agricultural Development and Advisory Service.* Ministry of Agriculture, Fisheries and Food, Pinner.

Nichols, D. (ed.) (1983). *Safety in Biological Fieldwork — Guidance Notes for Codes of Practice*, 2nd edn. Institute of Biology, London.

Pearce, M.E. (1984). *Medicines and Poisons Guide*, 4th edn. Pharmaceutical Press, London.

Post Office (annual). *Post Office Guide.* Post Office, London.

Rogers, W.V.H. and Clarke, M.G. (1978). *The Unfair Contract Terms Act 1977.* Sweet & Maxwell, London.

Seamer, J.H. and Wood, M. (1981). Codes of Practice. In *Safety in the Animal House*, Laboratory Animals Handbook 5 (Seamer, J.H. and Wood, M., eds), 2nd edn. Laboratory Animals, London.

Webster, C.A.R. (1981). *Environmental Health Law.* Sweet & Maxwell, London.

RECOMMENDED READING

ABPI (1982). *The Safe Storage and Handling of Animal Medicines*. Association of the British Pharmaceutical Industry, London.

Cooke, A.J.D. (1976). *A Guide to Laboratory Law*. Butterworth, London.

Cooper, M.E. (1981). Legal requirements. In *Safety in the Animal House, Laboratory Animals Handbook 5* (Seamer, J.H. and Wood, M., eds), 2nd edn. Laboratory Animals, London.

Dewis, M. (1978). *The Law on Health and Safety at Work*. Macdonald and Evans, Estover, Plymouth.

HSC (1982). *The Safe Disposal of Clinical Waste*. Health and Safety Commission/Health Services Advisory Committee. HMSO, London.

Mitchell, E. (1977). *The Employer's Guide to the Law on Health, Safety and Welfare at Work*, 2nd edn. Business Books, London.

Nichols, D. (ed.) (1983). *Safety in Biological Fieldwork — Guidance Notes for Codes of Practice*, 2nd edn. Institute of Biology, London.

Seamer, J.H. and Wood, M.(eds) (1981). *Safety in the Animal House*, Laboratory Animals Handbook 5, 2nd edn. Laboratory Animals, London.

Selwyn, N. (1982). *Law of Health and Safety at Work*. Butterworth, London.

Webster, C.A.R. (1981). *Environmental Health Law*. Sweet & Maxwell, London.

Wheelton, R. (1977). *A Study of the Arrangements for Radiological Protection in Twenty-three Veterinary Practices in Scotland*. National Radiation Protection Board, Harwell.

Health and Safety Commission Leaflets:

HSC 2 *The Health and Safety at Work etc. Act 1974: The Act Outlined*
HSC 3 *The Health and Safety at Work etc. Act 1974: Advice to Employers*
HSC 4 *The Health and Safety at Work etc. Act 1974: Advice to the Self-employed*
HSC 5 *The Health and Safety at Work etc. Act 1974: Advice to Employees*
HSC 6 *Guidance Notes on Employers' Policy Statements for Health and Safety at Work*
HSC 7 *Regulations, Codes of Practice and Guidance Literature*
HSC 8 *Safety Committees: Guidance to Employers whose Employees are Not Members of Recognised Independent Trade Unions*
HSC 9 *Time Off for the Training of Safety Representatives*

Health and Safety Executive Leaflets:

HSE 4 *Short Guide to the Employers' Liability (Compulsory Insurance) Act 1969*
HSE10 *Library and Information Services*
HSE13 *Directorate of Information and Advisory Services*
HSEL1 *HSELINE: the New Computerised Source of Health and Safety at Work References, 1981*

9 *Foreign and International Legislation Relating to Animals*

If a bird's nest chance to be before thee in the way in any tree, or on the ground, whether they be young ones, or eggs, and the dam sitting upon the young, or upon the eggs, thou shalt not take the dam with the young.

(Deuteronomy 22 v.6)

'O Prophet! I passed through a wood and heard the voices of young birds and I put them in my carpet.' Then the Prophet said '. . . Return them to the place from which you took them and let their mother be with them.'

(From the sayings of the Prophet Mohammed)

Various aspects of the law relating to animals have so far been examined in respect of Great Britain. Many other nations have comparable provisions, although there may be considerable variation in terms of scope, usage and enforcement. This final chapter provides an introduction to other countries' law of animals with particular focus on North America and Western Europe.

It would require a volume in itself to provide a summary of all relevant legislation for every country, besides which, such information is not always readily available. The purpose of this chapter is to provide an introduction to the appropriate legislation and to provide sources, particularly in terms of references and useful addresses (see Appendix 1), which the reader may pursue as required.

While the provisions and the means of implementation may vary, it is reasonable to expect that, in any country, some form of regulation exists — the problem may lie in obtaining the information, understanding its impact or keeping abreast of amendments. The primary source of information on national law is normally government departments and copies of legislation can usually be obtained from official printers, although the diplomatic mission of the country involved may be able to provide or obtain data. Voluntary organisations at home or overseas, as well as the legal profession in the country involved, may also be useful sources of information on legislation.

INTERNATIONAL LEGISLATION

International legislation, in the form of treaties and conventions, may be made between individual countries, by many countries joining a worldwide agreement (such as CITES) or by regional groups of countries, such as the EEC. Such laws, with some exceptions, have no effect until their provisions have been incorporated in the national law of the countries which are party to them. Bringing a treaty into effect depends upon signature by the participating countries, ratification by them and, finally, implementation. This may involve wholesale incorporation of the treaty's provisions, as for the Belgian Law approving the European Convention for the Protection of Animals in International Transport 1971, or automatic incorporation into national law, as with EEC regulations; alternatively, the amendment of existing legislation or the making of totally new laws may achieve the desired effect.

At times, the preparation of an international convention arises from a general movement to update legislation on a particular subject or *vice versa*. Thus, in the 1970s many European countries produced or revised their legislation relating to animal welfare and conservation. For example, the Council of Europe Convention for the Protection of Vertebrate Animals used for Experimental and other Scientific Purposes was in preparation for many years until it came into force in 1986. Its implementation depends upon its ratification and the provision of national laws by individual countries (see Chapter 4). Great Britain has revised its legislation in the light of the Convention, the EEC Directive on the subject and current thinking at national level (see Chapter 4); other countries have revised or introduced legislation on the subject in the recent past. A similar worldwide development in national law can be observed in the case of the CITES Convention (see later and Chapter 7).

In many cases a convention or treaty provides minimum standards for legislation and it is not uncommon for countries to maintain more stringent or additional provisions in some areas. For example, import and export of birds of prey between the UK and Europe are more heavily restricted than is required by the appropriate EEC regulations. Conversely, a country may also expressly derogate from treaty provisions which it declines to implement.

The EEC was established by the EEC Treaty (Treaty of Rome) which provides for several types of legislative provision, namely regulations, directives, decisions, treaty provisions and rules, recommendations and opinions. Of these, the most frequently encountered are the first three. Regulations, directions and decisions may be made either by the Commission or the Community. Regulations, by virtue of Article 189(2) of the Treaty of Rome, take effect immediately, being automatically incorporated in a member country's national law. Most decisions, treaty provisions and rules, together with directions, have to be specifically enacted in national legislation before they become effective. Recommendations and opinions are not enforceable. EEC legislation and allied provisions and deliberations are published in the *Official Journal of the EEC*.

THE LAWS OF INDIVIDUAL COUNTRIES

Many countries make legal provisions relating to animals although the extent, format and degree of enforcement may vary greatly. In some cases the legislation is based on that of former imperial powers; thus, statutes modelled on British law are to be found in Third World countries such as Kenya and India and in old Commonwealth countries such as Australia and New Zealand. Indeed, as the quotation preceding Chapter 3 indicates, English emigrants to America produced what is thought to be the earliest recorded animal welfare legislation (Leavitt and Halverson, 1978). In Moslem states, reliance may be placed on principles of religious law rather than on specific legislation affecting animals (Ba Kader *et al.*, 1983/1403H; WSPA, undated; WWF, 1987). This chapter, when dealing with national legislation, will deal mainly with North American and Western European countries.

While copies of individual pieces of legislation are generally available, some countries or bodies publish volumes of their legislation on specific subjects relating to animals, e.g. Brazil (FBCN, 1983), France (Anon, 1984a; 1984b), and Spain (Anon, 1980). In addition, lists, summaries and collections which have been made for various purposes may provide a useful starting point for further information. Those relevant to animal law include the following:

(a) The *Information Bulletin on Legal Affairs within the Council of Europe and in Member States* which publishes a quarterly summary of developments in Council and national legislation

(b) Projects undertaken by the IUCN Environmental Law Centre in Bonn, and the IUCN Commission on Environmental Policy Law and Administration project.

These involve the collation of national and international environmental law and include an *Index to Species Dealt with in Legislation* which forms part of the data bank of the Centre's Environmental Law Information Service (Farrell, 1981; 1984)

(c) A worldwide list of national legislation giving titles of laws affecting chiefly wild animals and an indication of their content (Traffic, 1979)

(d) A summary of West and East European national legislation relating to reptiles and amphibians (Bruno, 1973). See Appendix 3 (Note 22).

(e) Surveys, with references to the law, relating to the protection of birds in Mediterranean countries (Woldhek, 1980) and to raptors in Europe (Conder, 1977) and elsewhere (Hilton, 1977. See also Chancellor and Meyburg (1986), Robinson (1987)) and IUCN (1986)

(f) Outlines of national legislation relating to animal research published in successive editions of *Primate Report* (Spiegel, 1978, *et seq.*), in Hampson (undated), Rankin (1984), Rowsell (1985), Anon (1987) and Tuffrey (1987)

(g) In the United States comparative collections of state laws on specific topics relating to animals.

These have been compiled, for example, by the Animal Welfare Institute (Leavitt, 1978a; McGaugh and Genoways, 1976); commentary on federal laws is to be found in Fox (1980), Leavitt (1978a) and Bean (1983)

(h) Summaries of European animal welfare legislation in Ray and Scott (1973) and Taylor (1975, 1977); a study of Australian, North American and European animal welfare laws (Hansard, 1984)

UNITED STATES OF AMERICA

The USA has several tiers of legislation. Federal laws apply throughout the United States; they are to be found in the US Code and are implemented by rules which are contained in the Code of Federal Regulations (Chaloux and Heppner, 1980). In addition, individual states make laws for application within their boundaries. There may also be more localised district or city laws and bye-laws. Favre and Loring (1983) consider a broad spectrum of federal and state statutes and case law relating to animals.

Federal law relating to animals may only be formulated in the fields in which Congress has been given legislative power by the US Constitution; this authority may be dependent upon extrapolation from more general authority such as Congress's treaty-making power or clauses of the constitution conferring legislative power in the areas of property and commerce. The extent and relationship of federal and state legislation have been extensively examined in respect of wildlife legislation by Bean (1983). A summary of federal animal law, enforcement agencies and sources of state laws is to be found in Fox (1980).

Generally, most animal legislation is to be found at state level. Such laws deal with, for example, the cruel treatment of animals, humane slaughter and the use of animals in sport and exhibition; there may be specific provisions for species such as dogs, cats and horses and for specific matters such as trade and transport. Wildlife is protected by state laws regulating the taking or keeping of listed species and the methods used for doing so, although authorisation may be obtained for activities which are otherwise forbidden. Useful summaries of state laws affecting animals are to be found in Leavitt (1978a) and McGaugh and Genoways (1976).

At federal level the main Acts which affect animals are as follows:

(a) The Animal Welfare Act 1966 (as amended 1970, 1976) deals with interstate movement and commercial dealing involving various species of animals. It forbids animal fights (although game bird fights are outlawed only where state laws have that effect). It also regulates the supply and care of animals destined for research facilities or exhibition in the pet trade (Stevens, 1978a; Chaloux and Heppner, 1980). The 1966 Act has been further amended by the Improved Standards for Laboratory Animals Act 1985 which requires the establishment of institutional animal committees and the provision by USDA of standards for the care of animals used in research

(b) The Twenty-eight Hour Law 1906 deals with interstate rail transport of animals destined for slaughter. It requires resting, feeding and watering periods and a restriction on transportation for more than 28 hours without such breaks. The importance of this Act has declined with the increase in road transportation

(c) The Humane Slaughter Act 1958 (as amended 1978) requires livestock (cattle, calves, horses, mules, sheep and swine) to be slaughtered by specified methods, including ritual slaughter. In addition, the Federal Meat Inspection Act 1958 (as amended 1979) provides for inspection of handling and slaughter methods for the foregoing species as well as other Equidae and goats, to ensure compliance with the Humane Slaughter Act (Leavitt, 1978b)

(d) The Horse Protection Act 1970 (as amended 1976) and the Wild Horses and Burros Act 1971 (as amended 1978) provide protection from various forms of cruelty and the capture and killing of wild horses which live on federally owned public land (Twynne, 1978; Bean, 1983)

(e) The Bald Eagle Protection Act 1940 (as amended 1959, 1962, 1972) protects the national emblem of the USA, the bald eagle, by making it illegal to take or possess such birds or golden eagles (which are not always readily distinguishable), their eggs or nests. Under pemit, they may be taken for scientific or educational purposes such as for museums and zoos (McGaugh and Genoways, 1976; Bean, 1983)

(f) The Lacey Act 1901 (as amended 1948, 1949, 1960, 1969) prohibits the interstate transport of wildlife which has been killed in violation of state laws

(g) The Migratory Bird Treaty Acts 1916–1976 are based on various treaties with Great Britain, Mexico and Japan and put restrictions on hunting, killing, taking and possession of birds protected by the treaties

(h) The Endangered Species Act 1969 (as amended 1973, 1978) implements CITES in the USA and deals not only with national and international commerce in the species listed by the US Department of the Interior as endangered or threatened but also with their protection from hunting, killing, taking and injuring. Protection is afforded to their habitat in that areas vital to their survival must not be put at risk by the activities of federal bodies (McGaugh and Genoways, 1976; Bean, 1983)

(i) The Marine Mammal Protection Act 1972 (as amended 1973, 1976) protects species such as dolphins, whales and seals in respect of killing, harassment, hunting and capturing. It requires permits to be obtained to authorise possession and sale. Some exceptions are made for traditional hunting by indigenous populations (Stevens, 1978b)

EUROPE

In western and eastern Europe individual countries have a variety of legislation affecting animals, some of which are referred to elsewhere in this chapter.

OTHER COUNTRIES

Again, many countries have specific laws or parts of codes of law relating to animals, particularly if they are derived from the British or Roman law and continental European legal systems. In other countries principles may derive from religious laws (see earlier).

In the British Commonwealth countries, aspects of the law affecting animals in New Zealand have been discussed by Morgan (1967) and as to animal welfare by Wells (1983). For Australia, Blood (1985) covers the federal and state laws and Hansard (1984) examines the animal welfare provisions.

NATIONAL LEGISLATION RELATING TO SPECIFIC TOPICS

Welfare Legislation

USA

As mentioned earlier, in the USA much of the legislation dealing with the welfare of animals is made by individual states. This can lead to variation from state to state in respect of requirements and methods of enforcement. Welfare law relevant to cruel treatment, humane slaughter, organised animal fights and sports, trapping and poisoning have been summarised and compared in Leavitt (1978a). There is also federal legislation, particularly in respect of proper care in transportation and commerce. Burr (1975) has reviewed anti-cruelty law in the USA and produced a model statute for animal welfare. Favre and Loring (1983) have examined the US legislation and problems in its interpretation.

Europe

Most western European countries have animal welfare legislation dealing with cruelty to individual animals and in exhibition, sport and commerce. The legislation of EEC countries has been summarised by Taylor (1975) and discussed by Deleuran (1977). The Council of Europe has produced three relevant conventions which have been widely adopted amongst its members and by some other countries (Simonsen, 1982; Wiederkehr, 1982). They are:

Convention for the Protection of Animals during International Transport 1968
Convention for the Protection of Animals for Slaughter 1972
Convention for the Protection of Animals kept for Farming Purposes 1976 (see Appendix 3 (Note 23))

The first and last mentioned Conventions have also been approved by the Council of the EEC and it has also made a council directive on the stunning of animals before slaughter (EEC, 1974).

The Council of Europe *ad hoc* Committee of Experts for the Protection of Pet Animals is also preparing a Convention on the Protection of Pet Animals and a Convention on the Protection of Animals in Exhibition Situations.

The problems of implementing national and EEC legislation illustrated by the RSPCA's complaint to the Commission regarding the transport of live cattle and poultry from the UK to France in breach of such law (Muriel *et al.*, 1985). See Appendix 3 (Note 24).

Some examples of individual countries' animal welfare follow.

Belgium

The Protection of Animals Act 1975 makes it an offence to treat an animal cruelly or badly or to cause death, wounding, mutilation or suffering, to fail to give an animal necessary care or to abandon or overwork it, to infringe the humane slaughter regulations, to buy or sell a blind bird or to organise animal fights. There is provision for the confiscation of animals which are the subject of such offences. See Appendix 3 (Note 25).

Norway

Under the Welfare of Animals Act 1974 a person responsible for a domestic animal must give it proper care and living conditions. There is a duty to help a sick or injured animal or, alternatively, to inform the police in the case of a domestic animal, tame reindeer or large game animal; those animals *in extremis* may be killed humanely. The Act lists forms of cruelty, such as beating, which are forbidden. An animal must be killed humanely. Public exhibits of animals must be licensed by the Ministry of Agriculture, as must some forms of commerce such as boarding establishments. Pet shops and other trade in animals or the hiring of horses require the permission of a County Veterinary Officer. The Act is enforced by Animal Welfare Boards which are appointed by each local authority and which include a Ministry of Agriculture veterinary surgeon.

France

Animal welfare legislation is published in an official collection (Anon, 1984a). Law No. 76–629 of 1976 makes cruelty to, and the overworking of, domestic and captive wild animals an offence.

Denmark

By the Law on the Protection of Animals 1950 animals must be treated properly and not subjected to unnecessary suffering. They must be given proper living conditions. The Act forbids many specific forms of cruelty; slaughter must be humane; licences are required for pet shops, animal exhibits and zoos; performing animals must be trained humanely. Treatment of animals must be by a veterinary surgeon except in an emergency or when only momentary suffering is caused. Castration must normally be performed under anaesthesia by a veterinary surgeon, as must tail docking of horses and cows and dogs and the cropping of dogs' ears, such procedures being permitted only for health reasons.

Switzerland

The Confederate Animal Protection Act 1978 and Confederate Animal Protection Order 1981 deal with all aspects of animal welfare and are examined in detail by Goetschel (1986). See Appendix 3 (Note 26).

Other Countries

Many other countries have some form of animal welfare legislation, such as the Animals and Birds Act 1970 of Singapore and the Prevention of Cruelty to Animals Act 1957 of Mauritius.

The relevant laws of New Zealand have been considered by Wells (1983); those of Australia have been examined in conjunction with other legal systems by the Select Committee of the Commonwealth of Australia Senate (Hansard, 1984) and the laws of each state or territory are also listed in the tables of legislation provided by Blood (1985). The South African cruelty laws are discussed briefly by Austin (1985).

Animals Used in Research

USA

The Animal Welfare Act 1966 (as amended) (see earlier) controls the supply and care of animals (defined as cats, dogs, non-human primates, rabbits, hamsters and guinea pigs) destined for, and kept in, research facilities. The Act does not apply to rats and mice although these and other warm-blooded species could be designated as subject to the Act. Facilities must obtain their animals from licensed dealers, mark such cats and dogs and keep records of the latter (Stevens, 1978a).

The Act expressly does not apply to the design and performance of experiments using animals. The protection of such animals and the regulation of experimental procedures are dependent, not on legislative provisions, but on various administrative structures, inspection and financial control by grant-giving agencies. However, the Act provides that research facilities must meet certain standards in the care, treatment and use of animals undergoing experiment including the use of anaesthesia and analgesia. The Health Research Extension Act 1985 provides for guidelines on animal usage in research in biomedical and behavioural sciences.

Research facilities must be registered with the US Department of Agriculture (USDA) and an annual report relating to animal usage and the administration of anaesthetics and analgesics must be submitted. The responsible institution must have an animal care committee and the research facility is subject to inspection by USDA.

Federal grant-giving bodies, in particular the National Institutes of Health

(NIH) of the Public Health Service of the US Department of Health and Human Services, are expected to exercise some control over the care of animals used in the experimental work which they fund.

The NIH require research facilities to comply with its *Guide to the Care and Use of Laboratory Animals* which lays down standards for housing, hygiene and the use of anaesthetics and analgesics. In addition, the facility must have a committee to screen and monitor research projects with a view to animal welfare. The committee must report to the Office for Protection of Research Risks on compliance with the Guide and inspections may be made by NIH officials.

If the research facility does not have such a committee it is required to be accredited by the American Association for Accreditation of Laboratory Animal Care (AAALAC) which sets standards and inspects research facilities (Ramsay and Spinelli, 1982; Dodds, 1984).

Pressure for the maintenance of standards of animal care also come from the Science Education Administration of USDA which funds agricultural research (Soave and Crawford, 1981) and the Food and Drug Administration (FDA) which requires the laboratory testing for safety of drugs, food additives and biological and other products in accordance with its standards of Good Laboratory Practice. While the latter relates largely to quality control it has an influence on standards of animal care and is applied not only in the USA but also to research facilities in other countries which produce toxicological data to be submitted for FDA approval (Meyers, 1983). The Food Security Act 1985 promotes the use of alternatives to animals. The foregoing requirements for research facilities, together with other legislation which affects the use of animals in experiments in the USA, are discussed in Johnson and Morin (1983), Meyers (1983) and Schwindaman (1983).

Canada

In Canada, with the exception of two provinces, the control over experimental animals is exercised by self-regulation on the part of research institutions rather than by legislation.

The Canadian Council on Animal Care (CCAC) has provided standards and guidelines and a structure for self-regulation by internal animal care committees backed up by site visits made by CCAC-appointed assessment panels. The CCAC's *Guide to the Care and Use of Experimental Animals*, Volume 1 (CCAC, 1980), provides a structure of ethical requirements (e.g. in respect of pain), together with standards and guidance on housing, environment, care and experimental usage. Volume 2 (CCAC, 1984) applies such information to a wide variety of species used in research. The application of this system has been discussed by Rowsell (1983, 1984).

The provinces of Alberta and Ontario have legislation relating to animals used in experiments. They and Saskatchewan also have legal provisions relating to the procurement of cats and dogs for research (CCAC, 1980; Rowsell, 1974).

Europe

Most Western European countries have legislation relating to the use of animals in research, and these have recently been summarised by Hampson (undated) and Rowsell (1985). Rankin (1984) has discussed these laws briefly and summaries have been published in *Primate Report* (Spiegel, 1978, *et seq.*); Öbrink (1982, 1984) has described the Swedish provisions and Goetschel (1986) has examined the Swiss law. However, in view of the Council of Europe Convention for the Protection of Vertebrate Animals used for Experimental and other Scientific Purposes 1986 (see Chapter 4 and Hovell (1985)) it is likely that many of these national laws will be subject to amendment in the near future. The EEC Directive (EEC, 1986) requires that by 1988 member countries should conform with its provisions for the control of experiments carried out for the testing of products and substances and for the purpose of environmental protection.

Eastern European countries such as Poland (described by Radzikowksi (1984)), Hungary and Czechoslovakia have such legislation (Rowsell, 1985). Some examples of individual national legislation are given below (see also Appendix 3 (Note 27)).

France

The Decree Regulating Experiments and Scientific and Experimental Research involving Live Animals 1968 requires that experiments are performed by a licensed person or under the supervision of the licensee. Invertebrates, observational field studies of wild animals and non-painful experiments are not covered by the Act.

Animals destined for research must be properly housed and cared for and unnecessary suffering avoided. Records of the source of animals must be kept for one year and be available for inspection. Licensing, inspection and enforcement of any particular establishment lies with the government minister responsible for that body.

Only painful experiments are subject to the Act and it requires that anaesthesia or analgesia should be used unless they are incompatible with the needs of the research involved. An animal may not be used for more than one operation without anaesthesia except in justifiable cases of unavoidable necessity. Either an anaesthetised animal should be killed before anaesthesia has worn off or, if it is allowed to recover, every effort should be made to provide due care and to spare it post-operative suffering.

Federal Republic of Germany

The Animal Protection Act 1972, section 5, provides that either organisations or individuals performing experiments must hold permits to do so. Employees of an establishment may carry out animal research, as may others using its facilities, with the consent of the director.

A permit may be issued, when there is no alternative to the use of animals, for purposes of treatment and prevention of disease and scientific research. Proper care and treatment of experimental animals must be provided. The number of experiments must be minimised and warm-blooded species and higher orders may only be used if others are not adequate for the proposed research. So far as is compatible with the purpose of an experiment, pain, suffering and harm must be avoided, anaesthesia must be used and painful operations restricted to one per animal. Survival after the conclusion of an experiment is decided by a veterinarian or the researcher, depending upon the species involved. Records of experiments must be kept and are subject to inspection. See Appendix 3 (Note 28).

Belgium

Under the Protection of Animals Act 1975 it is an offence to carry out experiments on living animals unless the experiments are necessary for the purposes of scientific research, warfare or veterinary science. They must be performed in university laboratories under the control of the director responsible for the laboratory concerned. Anaesthesia (unless otherwise dictated by the experiment) must be used and post-operative care must be provided. Veterinary inspectors are responsible for monitoring these provisions and have access at all times to laboratories. See Appendix 3 (Note 29).

Other countries

Outside Europe, many countries which were at one time governed by Great Britain (e.g. Kenya, Jamaica) still have legislation similar to the 1876 Act on their statute books, athough it may not now be enforced. Australia is currently reconsidering its welfare law and the State of New South Wales is preparing legislation providing for self-regulation in animal research (Rose, 1984).

Some countries have no provisions at all (Abdussalam, 1984) or rely on religious principles. Austin (1985) states that in South Africa the welfare of laboratory animals is covered only by the general provision against cruelty in the Animals Protection Act 1962. See also Schoebaum and Benhar (1985).

The Appendix to Leavitt (1978a) lists countries which have relevant legislation and *Laboratory Animal Legislation* (UFAW, 1986) is a list of national laws in this field.

In addition to legislation many countries, organisations and institutions have codes of practice for the use of laboratory animals and an international code has been produced by CIOMS (CIOMS, 1985; Howard-Jones, 1985) and IASP (Zimmermann, 1984).

Animal Health

Matters of animal health are normally the responsibility of the state veterinary service of individual countries and are of such importance that almost every country must have some form of legal control. In the USA the Animal and Plant Health Inspection Service (APHIS) of USDA is responsible for notifiable diseases and their control, except when such duties are carried out by accredited private veterinarians, although there is also extensive state supervision (Soave and Crawford, 1981).

Numerous provisions have been made by the EEC for interstate control and eradication of disease and other matters relating to farm livestock (Taylor, 1972; Anon, 1975a).

Importation and exportation of animals are strictly controlled in most countries and may be comparable to the British system described in Chapter 5. Owing to the variability of disease prevalence and outbreaks, the requirements of importing countries differ considerably not only between countries (or states in larger countries) but from time to time within any one country. It is therefore necessary to obtain up-to-date information from the relevant diplomatic mission or ministry of agriculture and by reading the veterinary press.

Treatment and Care of Animals

USA

In the USA the veterinary profession is regulated through state legislation. Generally, a veterinary surgeon must be licensed by any state in which he wishes to practise; this will also involve passing an examination and providing proof of veterinary medical training. The legislation implemented by boards of veterinary medical examiners restricts the right of practice to veterinary surgeons. However, the legislation is such that this right extends only to domestic animals and the general scope may vary from state to state. Common exemptions from the need to be registered as a veterinary surgeon include the treatment by owners of their animals and public service and military veterinary surgeons (Soave and Crawford, 1981). Other aspects of law are also examined in Soave and Crawford (1981) and include animal welfare legislation, relevant principles of contract, negligence and nuisance and federal controls affecting veterinary practice. Ethical matters are handled by local state and national professional associations, the Judicial Council of the American Veterinary Medical Association being the highest level. Professional obligations are set out in *The Principles of Veterinary Medical Ethics* (Roberts, 1976; Soave and Crawford, 1981).

Europe

Each country controls the practice of veterinary surgeons within its borders. The EEC Directives on the Veterinary Profession provide for the recognition between EEC members of veterinary qualifications and of the right of holders of such qualifications to practise in EEC countries, subject to registration by the country where a veterinarian proposes to practise (Porter, 1979). Two recent cases in the European Court of Justice on the implementation of the EEC Directives have held that an individual veterinary surgeon may rely on the Directive even if it is not implemented in a particular EEC country and if registration is withheld in breach of the EEC Directive. Thus, a Dutch national who qualified in Holland but practised in Italy was refused registration in Italy on the grounds that Italy had not implemented the Directive. He was then prosecuted in an Italian court for unauthorised practice. The European Court, in giving its interpretation of the Directive at the request of the Italian court, ruled that Italy could not enforce its legislation prohibiting unregistered practice when it was itself in breach of the Directive; the veterinary surgeon could rely on the Directive even if it was not implemented (Rienks [1985]).

The European court followed its earlier decision in the similar case of Auer (Vincent) *v*. Ministère Publique [1985] in which it held that a national was entitled to practise in his own country, France, holding an Austrian qualification certified in Austria as complying with the Directive, although France had not implemented the latter and refused to register M. Auer.

Two "Eurovet" books (BVA, 1972; Anon, 1975a) describe the background and structure of practice and the state veterinary services in Denmark, Belgium, Luxembourg, the UK, France and the Federal Republic of Germany. They also deal with the EEC legislation relevant to veterinarians.

In Holland the practice of veterinary surgery is controlled by the Veterinary Medicine (Practice) Act 1954 (as amended, but see below). It reserves the right to practise to veterinary surgeons qualified in The Netherlands, or, with consent, elsewhere. The Act defines the practice of veterinary medicine as "giving medical, surgical or obstetric advice or assistance in respect of animals, operating on healthy animals, vaccinating animals and administering local or general anaesthetics to animals in a professional capacity" (Anon, undated). Farmers may (subject to limitations, for example, as to castration) treat and vaccinate their own livestock. There are also clearly defined exceptions for students and for those authorised to vaccinate poultry and castrate domestic species. The structure and work of the Dutch veterinary profession and state veterinary service are described in Anon (undated). Dutch veterinary legislation on practice, medicines, disease control and animal welfare is under revision (Janssen, 1985).

The Federation of Veterinarians in Europe has adopted a code of conduct as a model of minimum requirements on which veterinary professions of member countries may base their national codes; however, many countries have established more stringent requirements (FVE, 1981).

Other countries

Veterinary services provided by the state and privately are regulated in most countries. In the Third World veterinary auxiliaries may provide basic services, particularly in rural areas. Animal welfare societies may supplement these.

The law governing veterinary practice in Australia is to be found in Blood (1985). Based on the law of the State of Victoria, the book gives comparative tables for the corresponding statutes of other States. Most aspects of veterinary practice and ethics are also discussed with reference to the relevant UK and USA law.

Ignace (1984) has described the history and work of the veterinary profession in Mauritius.

Conservation

International

One of the areas where extensive international legislation has been produced is in the conservation of wildlife. In its turn, it has had worldwide influence on national wildlife legislation. See Appendix 3 (Note 29).

International agreements

There are four international treaties of global significance for wildlife conservation. They are:

Convention on International Trade in Endangered Species of Wild Fauna and Flora 1973 (CITES or Washington Convention)

Convention on Wetlands of International Importance especially as Waterfowl Habitat 1971 (Ramsar Convention)

Convention concerning the Protection of the World Cultural and Natural Heritage 1972 (World Heritage Convention)
Convention on the Conservation of Migratory Species of Wild Animals 1979 (Bonn Convention) (see Appendix 3 (Note 30))

These four Conventions underpin the World Conservation Strategy (IUCN, 1980; NCC, 1983; 1984). They are the means by which it and the World Charter for Nature may eventually be given legal effect. The Charter is considered "soft law" in that it is not binding upon individual countries although having been adopted by the United Nations General Assembly there is a moral obligation to comply with it (Farrell, 1984; Lyster, 1985).

Other relevant worldwide conventions are the International Convention for the Regulation of Whaling 1946 and the Law of the Sea Convention 1982. The major international conventions are reproduced by Lyster (1985) and discussed by Kiss (1976) and Lyster (1985). See Appendix 3 (Note 31).

Regional organisations or areas also have interstate conservation agreements, e.g. the African Convention on the Conservation of Nature and Natural Resources 1968, the Convention for the Conservation of Antarctic Seals 1972, the Convention for the Conservation of the Vicuna 1969, and conventions for the Red Sea, Caribbean and (in preparation) East African regions (Lyster, 1985). Treaties may also be concluded between individual countries, such as the Migratory Birds Treaty made between the USA, Canada and the UK.

The Antarctic is subject to international legislation and agreements for the conservation and preservation of the area south of 60° south latitude (Edwards and Heap, 1981; Auburn, 1982; Couratier, 1983; Bonner, 1985; Lyster, 1985; Laws, 1986).

The most effective of the main conservation conventions is CITES, which relates to the international movement of endangered species (see Chapter 7). National CITES management authorities are required to keep records of imports and exports; an annual statistical report on trade in the UK and its dependent territories, including Hong Kong, is published by the DOE (DOE, undated; 1982). The Convention is extensively monitored by welfare societies and by the IUCN Wildlife Trade Monitoring Unit (Inskipp and Thomas, 1976). The implications of the environmental, trade and legal provisions are reported in international conservation publications such as WTMU (continuing) and IUCN (continuing).

USA

In the USA wildlife protection involves federal as well as state legislation. Most protection is provided at state level and can vary widely, depending upon the species at risk in the particular area and public attitudes to conservation; for example, the practice of falconry is forbidden in some states, permitted in others and tightly controlled in yet others, and there are federal regulations setting out standards for falconry permits in the two last cases. There are often exceptions to controls for indigenous peoples' traditional hunting rights (Bean, 1985).

The US federal legislation is discussed by Bean (1983) and Leavitt (1978a), and McGaugh and Genoways (1976) have collated the state laws which provide for the collection of protected species for scientific purposes. Fox (1980) gives the main federal laws together with the appropriate enforcement agencies. Controls on the keeping of wild animals as pets are discussed by Diesch (1981).

Europe

Within Europe, the Berne Convention (the Council of Europe Convention on the Conservation of European Wildlife and Natural Habitats 1979), the EEC Directive on the Conservation of Wild Birds and the EEC accession to CITES in its own right have influenced the law of western European countries. Their legislation protecting the wild species is likely to be based on the standards set by the Berne Convention, restricting the killing, taking and injuring of specified wild animals and regulating the methods which may be used to kill or take any wild species. Individual countries' legislation may vary in form, the species protected (which will depend on their status in that area) and the activities which are controlled since signatories may provide more extensive provisions than those laid down by the Convention.

The effect of the EEC acceding to CITES is such that member countries have two systems of trade control to administer. Movement within the EEC is based upon EEC documentation recognised by member countries, thus relieving a dealer from the need to complete new papers for each transfer between EEC countries. However, animals or derivatives first entering the EEC require the CITES international documentation, although on subsequent transfers within the EEC such documents are recognised by other EEC states. The structure is explained in DOE (1987), with tables showing the paperwork required for particular transactions.

Some examples of national legislation for the protection of wildlife which have been published as a collection or survey are available for Europe (Conder, 1977), France (Anon, 1984b), Spain (including the Balearic Islands) (Anon, 1980), Portugal (Anon, 1974) and Luxembourg (Anon, 1975b). See Appendix 3 (Note 32).

Other countries

The majority of countries outside Europe also have conservation legislation whereby individual species are protected in certain areas or all species in a particular area are protected. As mentioned earlier, Traffic (1979) includes references to such statutes. The IUCN Environmental Law Centre has published a worldwide index to wildlife species legislation and has prepared computerised data in this field (IUCN, 1977; Farrell, 1984). The IUCN Law and Policy Paper Series is devoted to studies of fundamental issues of national and international conservation legislation (Farrell, 1981; 1984; Ba Kader *et al.*, 1983; IUCN, 1987) (see Appendix 3 (Note 33)). The conservation laws of Brazil are published in FBCN (1983); see also IUCN (1986).

Owing to the growing interest in animal welfare and conservation the law on such subjects has expanded considerably since the early 1970s. Particularly in the field of wildlife, the drive towards international regulation has encouraged the collection and collation of national and regional laws on the subject.

Considerable difficulty in obtaining details of such legislation is often experienced and, since the information tends to be highly specialised and diverse, this chapter, while giving a guide to the existing legislation, is intended to lead the reader to sources from which more specific material may be acquired and from which the current state of law may be ascertained.

CASES

Auer (Vincent) *v*. Ministère Publique (No. 2) (No. 271/82) [1985] 1 C.M.L.R. 128, European Court.

Rienks (H.G.) (No. 5/83) [1985] 1 C.M.L.R. 144, European Court.

REFERENCES

Abdussalam, M. (1984). Animals for biomedical research: perspective for developing countries. In *Biomedical Research Involving Animals. Proposed International Guiding Principles* (Bankowski, Z. and Howard-Jones, N., eds). Council for International Organizations of Medical Sciences, Geneva.

Anon (undated). *Veterinary Work in the Netherlands*. Ministry of Agriculture and Fisheries, The Netherlands.

Anon (1974). *Lei da Caça, 1974*. Diario do Governo, 14 August 1974.

Anon (1975a). *Eurovet — 2*. Animal Health Services Division, Henderson Group One, London.

Anon (1975b). *Luxembourg geschützte Tiere*. Luxemburger Liga für Natur- und Umwetschutz. Graphic Centre Bourg-Bourger, Luxembourg.

Anon (1980). *Ley y Reglamento de Caza*. Boletin Oficial del Estado, Madrid.

Anon (1984a). *Protection de l'Animal*. Journal Officiel de la Republique Française. Brochure No. 1530 et Suppléments 1-3, Direction des Journaux Officiels, Paris.

Anon (1984b). *Protection des Especes de Faune et de Fleurs Sauvages*. Journal Officiel de la Republique Française, Textes Generaux No. 1454 — II. Direction des Journaux Officiels, Paris.

Anon (1987). Legislation and laboratory animals. In *The UFAW Handbook on the Care and Management of Laboratory Animals* (Poole, T., ed.). Longman, London.

Auburn, F.M. (1982). *Antarctic Law and Politics*. Hurst, London.

Austin, J.C. (1985). Legal requirements for the use of wild and exotic animals for teaching and research purposes in South Africa. In *Exotic Animals in Research* (Dawson, P., ed.). South African Association for Laboratory Animal Science, Pretoria.

Ba Kader, A.B.A., Al Sabbagh, A.L.T. El S., Al Glemid, M.H.S. and Izzidien, M.Y.S. (1983/1403H). *Islamic Principles for the Conservation of the Natural Environment*. IUCN Environmental Policy and Law Paper No. 20., IUCN, Gland, and Natural Resources/Kingdom of Saudi Arabia Meteorology and Environmental Protection Agency.

Bean, M. (1983). *The Evolution of National Wildlife Law*, 2nd edn. Praeger Press, New York.

Blood, D.C. (1985). *Veterinary Law, Ethics, Etiquette and Convention*. The Law Book Company, Sydney.

Bonner, W.N. (1985). Conserving the Antarctic. *Biologist* **32**, 145–150.

Bruno, S. (1973). *Problemi di Conservazione nel Campo dell' Erpetologia*. In Alli del III Simposio Nazionale salla Conservazione della Natura Vol. II. Cacucci Editore, Bari.

Burr, S.I. (1975). Toward legal rights for animals. *Environmental Affairs* **IV** (2), 205–254 (Boston College Environmental Law Center, Boston).

BVA (1972). *An Anatomy of Veterinary Europe*. Eurovet, London.

CCAC (1980). *Guide to the Care and Use of Experimental Animals*, Vol. 1. Canadian Council on Animal Care, Ottawa.

CCAC (1984). *Guide to the Care and Use of Experimental Animals*, Vol. 2. Canadian Council on Animal Care, Ottawa.

Chaloux, P.A. and Heppner, M.B. (1980). History and development of federal animal welfare regulations. *International Journal for the Study of Animal Problems* **1**(5), 287–295.

Chancellor R.C. and Meyburg, B.-U. (eds) (1986). *Birds of Prey Bulletin No. 3*. World Working Group on Birds of Prey and Owls of the International Council for Bird Preservation, London, Paris, Berlin.

CIOMS (1985). *International Guiding Principles for Biomedical Research Involving Animals*. Council for International Organizations of Medical Sciences, Geneva.

Conder, P. (1977). Legal status of birds of prey and owls in Europe. In *Report of Proceedings. World Conference of Birds of Prey, Vienna 1–3 October 1975*. (Chancellor, R.D., ed.). International Council for Bird Preservation, London.

Couratier, J. (1983). The regime for the conservation of Antarctica's living resources. In *Antarctic Resources Policy* (Vicuña, F.O., ed.). Cambridge University Press, Cambridge.

Deleuran, P. (1977). EEC legislation and the protection of domestic animals. *Animal Regulation Studies* **1**, 159–165.

Diesch, S.L. (1981). Should wild-exotic animals be banned as pets? *California Veterinarian* **12**, 13–18.

Dodds, J. (1984). Summary of Hopkins workshop (on animal care and use committees): papers and case presentations. *Newsletter* **6**(3), 1–10 (Scientists' Center for Animal Welfare, Washington, DC).

DOE (undated). *Report for 1981 on the Implementation in the United Kingdom of the CITES*. Department of the Environment, Bristol.

DOE (1982). *Report for 1982 on the Implementation in the United Kingdom of the CITES*. Department of the Environment, Bristol.

DOE (1987). *Controls on the Import and Export of Endangered and Vulnerable Species*, with Supplementary Notices 1–6. Department of the Environment, Bristol.

EEC (1974). Council Directive 74/577/EEC of 18 November 1984 on stunning of animals before slaughter. *Official Journal L316*, 10.

EEC (1986). Council Directive 86/609/EEC of 24 November 1986 on the approximation of laws, regulations and administrative provisions of the Member States regarding the protection of animals used for experimental and other scientific purposes. *Official Journal of the European Communities*, Vol. 29, No. L358 of 18 December 1986. Council of Ministers of the European Communities, Brussels.

Edwards, D.M. and Heap, J.A. (1981). Convention on the conservation of Antarctic marine-living resources: a commentary. *Polar Record* **20**, 353–362.

Farrell, A. (ed.) (1981). *Yearbook 1980–81*. World Wildlife Fund International, Gland.

Farrell, A. (ed.) (1984). *WWF Yearbook 1983/4*. World Wildlife Fund/IUCN Joint Information and Education Division, Gland.

Favre, D.S.F. and Loring, M. (1983). *Animal Law*. Greenwood Press, Westport.

FBCN (1983). *Legislaçao de Conservacáo de Natureza*. Fundacáo Brasiléira Para a Conservaçáo da Natureza. Companhia Energética de São Paulo, São Paulo.

Fox, M.W. (1980). *Returning to Eden*. The Viking Press, New York.

FVE (1981). *Code of Professional Conduct for Veterinarians in Europe*. Federation of Veterinarians in Europe, Brussels.

Goetschel, A.F. (1986). *Kommentar zum Eidgenössischen Tierschutzgesetz*. Verlag Paul Haupt, Bern.

Hampson (undated). *Laboratory Animal Protection Laws in Europe and North America*. Royal Society for the Protection of Animals, Horsham.

Hansard (1984). *Commonwealth of Australia Senate Select Committee on Animal Welfare*. Hansard, Canberra.

Hilton, J.R. (1977). The legal status of birds of prey in other parts of the world. In *Report of Proceedings. World Conference of Birds of Prey, Vienna 1–3 October 1975* (Chancellor, R.D., ed.). International Council for Bird Preservation, London.

Hovell, G.J.R. (1985). Council of Europe animal protection legislation — the role of ICLAS. In *The Contribution of Laboratory Animal Science to the Welfare of Man and Animals: Past, Present and Future* (Archibald, J., Ditchfield, J. and Rowsell, H.C., eds). Gustav Fischer, Stuttgart.

Howard-Jones, N. (1985). A CIOMS ethical code for animal experimentation. *WHO Chronicle* **39**, 51–56.

Ignace, J.A.E. (1984). *Two Centuries of Veterinary Profession in Mauritius 1771–1979*. Mauritius Society for the Prevention of Cruelty to Animals, Rose Hill.

Inskipp, T.P. and Thomas, G.T. (1976). *Airborne Birds. A Further Study of Importation of Wild Birds into the United Kingdom*. Royal Society for the Protection of Birds, Sandy.

IUCN (continuing). *IUCN Bulletin*, International Union for Conservation of Nature and Natural Resources, Gland.

IUCN (1977). *IUCN Environmental Law Centre Information Services*. IUCN Publications, International Union for Conservation of Nature and Natural Resources, Gland.

IUCN (1980). *World Conservation Strategy*. International Union for Conservation of Nature and Natural Resources, Gland.

IUCN (1986). *African Wildlife Laws* (de Klemm, C. and Lausche, B., eds). International Union for Conservation of Nature and Natural Resources, Gland and Cambridge.

IUCN (1987). *The IUCN Conservation Library Catalogue 1987*. International Union for the Conservation of Nature and Natural Resources, Gland.

Janssen, J. (1985). Significance of veterinary legislation ('*trias veterinaria*') for small animal practice. In *Voorjaarsdagen 1985*, Proceedings of Congress of the Royal Netherlands Veterinary Association and the Netherlands Small Animal Veterinary Association, Amsterdam.

Johnson, D.K. and Morin, L.M. (1983). US laws, regulations and policies important to managers of non-human primate colonies. *Journal of Medical Primatology* **12**, 233–238.

Kiss, A.C. (1976) *Survey of Current Developments in International Environmental Law*. International Union for Conservation of Nature and Natural Resources, Morges.

Laws, R.M. (1986). Animal conservation in the Antarctic. *Symposium on Advances in Animal Conservation*. Zoological Society of London, London.

Leavitt, E.S. (ed.) (1978a). *Animals and their Legal Rights*. Animal Welfare Institute, Washington DC.

Leavitt, E.S. (1978b). Humane slaughter laws. In *Animals and their Legal Rights* (Leavitt, E.S., ed.). Animal Welfare Institute, Washington DC.

Leavitt, E.S. and Halverson, D. (1978). The evolution of anti-cruelty laws in the United States. In *Animals and their Legal Rights* (Leavitt, E.S., ed.). Animal Welfare Institute, Washington DC.

Lyster, S. (1985). *International Wildlife Law*. Grotius Publications, Cambridge.

McGaugh, M.H. and Genoways, H.H. (1976). *State Laws as they Pertain to Scientific Collecting Permits*. Museology **2** (Texas Tech University, Texas).

Meyers, N.M. (1983). Government regulation of non-human primate facilities. *Journal of Medical Primatology* **12**, 169–183.

Morgan, E.D. (1967). *The Law of Animals*. Butterworth, Wellington.

Muriel, K., Mews, A. and Milner, F. (1985). *RSPCA Complaint to the Commission of the European Communities on the International Transport of Live Animals*. Royal Society for the Prevention of Cruelty to Animals, Horsham.

NCC (1983). *Ninth Annual Report*. Nature Conservancy Council, London.

NCC (1984). *Tenth Annual Report*. Nature Conservancy Council, London.

Obrink, K.J. (1982). Swedish law on laboratory animals. In *Scientific Perspectives on Animal Welfare* (Dodds, W.J. and Orlans, F.B., eds). Academic Press, New York.

Obrink, K.J. (1984). Monitoring of animal experimentation and ethical review committees. In *Biomedical Research Involving Animals. Proposed International Guiding Principles* (Bankowski, Z. and Howard-Jones, N., eds). Council for International Organizations of Medical Sciences, Geneva.

Porter, A.R.W. (1979). Veterinary qualifications and the right to practise in the EEC: an analysis of the directives. *Veterinary Record* **104**, 344–347.

Radzikowksi, C. (1984). In McCarthy, C.R., Monitoring animal experimentation: general considerations, Discussion. In *Biomedical Research Involving Animals. Proposed International Guiding Principles* (Bankowksi, Z. and Howard-Jones, N., eds). Council for International Organizations of Medical Sciences, Geneva.

Ramsay, D.J. and Spinelli, J.S. (1982). Responsibility of funding agencies: central or local? In *Scientific Perspectives on Animal Welfare*. (Dodds, W.J. and Orlans, F.B., eds). Academic Press, New York.

Rankin, J.D. (1984). Monitoring of animal experimentation: statutory controls. In *Biomedical Research Involving Animals. Proposed International Guiding Principles*. (Bankowski, Z. and Howard-Jones, N., eds). Council for International Organizations of Medical Sciences, Geneva.

Ray, P.M. and Scott, W.N. (1973). Animal legislation in the E.E.C. *British Veterinary Journal* **129**, 194–201.

Roberts, S.J. (1976). Reflections on the principles of professional ethics. *Journal of the American Veterinary Medical Association* **169**, 430–433.

Robinson, P.J. (1987). *Legal Status of Diurnal Birds of Prey in Africa*. Royal Society for the Protection of Birds, Sandy.

Rose, M. (1984). Personal communication.

Rowsell, H.C. (1974). Legislation regulations pertaining to laboratory animals — Canada. In *CRC Handbook of Laboratory Animal Science*. Vol. 1 (Melby Jr, E.C. and Altman, N.H., eds). CRC Press, Cleveland.

Rowsell, H.C. (1983). Monitoring of animal experimentation — codes of practice legislation or voluntary peer review. In *Biomedical Research Involving Animals*. Proposed International Guiding Principles (Bankowski, Z. and Howard-Jones, N., eds). Council for International Organizations of Medical Sciences, Geneva.

Rowsell, H.C. (1984). *The Animal in Research — Perspectives of Voluntary Control*. National Symposium on Imperatives in Research Animal Use, Washington. Canadian Council on Animal Care, Ottawa.

Rowsell, H.C. (1985). *A Comparative Overview of International Regulations, Guidelines and Policies on the Care and Use of Experimental Animals*. Canadian Council on Animal Care, Ottàwa.

Schoenbaum, M. and Benhar, E. (eds) (1985). *Ethical Aspects of the Use of Animals in Research*. Special issue. *Quarterly of the Israel Zootechnical Association* 14, 1–2.

Schwindaman, D.F. (1983). The Animal Welfare Act as applied to private animal laboratories. *Journal of Medical Primatology* **12**, 250–255.

Simonsen, H.B. (1982). Role of applied ethology in international work on farm animal welfare. *Veterinary Record* **111**, 341–342.

Soave, O. and Crawford, L.M. (1981). *Veterinary Medicine and the Law*. Williams and Wilkins, Baltimore.

Spiegel, A. (ed.) (1978, *et seq.*). *Primate Report*, **1** *et seq*. Verlag Erich Goltze, Göttingen.

Stevens, C. (1978a). Laboratory animal welfare. In *Animals and their Legal Rights* (Leavitt, E.S., ed.). Animal Welfare Institute, Washington DC.

Stevens, C. (1978b). Marine Mammals. In *Animals and their Legal Rights* (Leavitt, E.S., ed.). Animal Welfare Institute, Washington DC.

Taylor, G.B. (1972). EEC regulations and directives on notifiable diseases. In *An Anatomy of Veterinary Europe*. BVA, London.

Taylor, G.B. (1975). Animal welfare legislation in Europe. In *Animals and the Law*. Universities Federation for Animal Welfare, Potters Bar.

Taylor, G.B. (1977). Animal welfare legislation in Europe. *Animal Regulation Studies* **1**, 73–85.

Traffic (1979). *Legislation Catalogue* (First draft). Traffic (International), London (now WTMU, Cambridge).

Tuffrey, A.A. (1987). *Laboratory Animals: An Introduction for New Experimenters*. Wiley, Chichester.

Twynne, P. (1978). Horses. In *Animals and their Legal Rights* (Leavitt, E.S., ed.). Animal Welfare Institute, Washington DC.
UFAW (1986). *Laboratory Animal Legislation*. Universities Federation for Animal Welfare, Potters Bar.
Wells, N.E. (1983). The moral status of animals: reform of animal protection law. *Ph.D. thesis*, University of Auckland, Auckland.
Wiederkehr, M.-O. (1982). The Council of Europe's Conventions. In *Protection and Rights of Animals*. Forum Council of Europe 3/82, VII. Council of Europe, Strasbourg.
Woldhek, S. (1980). *Bird Killing in the Mediterranean*. European Committee for the Prevention of Mass Destruction of Migratory Birds, Zeist.
WSPA (undated). *Excerpts from the Islamic Teachings on Animal Welfare*. World Society for Protection of Animals, London.
WTMU (continuing). *Traffic Bulletin*. Wildlife Trade Monitoring Unit, Cambridge.
WWF (1987). Religion and Nature. Muslim Declaration. *Environmental Policy and Law* 17(1), 16, 47. Buddhist, Christian, Hindu and Jewish Declarations. *Environmental Policy and Law* 17(2), **69**, 87–90.
Zimmermann, M. (1984). Ethical considerations in relation to pain in animal experimentation. In *Biomedical Research Involving Animals* (Bankowski, Z. and Howard-Jones, N., eds). Council for International Organizations of Medical Sciences, Geneva.

RECOMMENDED READING

AATA (continuing) *Quarterly Newsletter*. Animal Air Transport Association, Fort Washington.
Anon (continuing). *Encyclopaedia of EEC Legislation*. Butterworth, London.
Anon (continuing). Title 7 *Agriculture* and Title 50 *Wildlife and Fisheries*. United States Code of Federal Regulations, U.S. Printing Office, Washington DC.
AWI (continuing). *The Animal Welfare Institute Quarterly*. Animal Welfare Institute, Washington DC.
Collins, L. (1975). *European Community Law in the United Kingdom*. Butterworth, London.
COE (1969). *Explanatory Report on the European Convention for the Protection of Animals during International Transit*. Council of Europe, Strasbourg.
COE (1976). *Explanatory Report on the European Convention on the Protection of Animals Kept For Farming Purposes*. Council of Europe, Strasbourg.
COE (1979a). *Explanatory Report Concerning the Convention on European Wildlife and Natural Habitats*. Council of Europe, Strasbourg.
COE (1979b). *Explanatory Report on the European Convention for the Protection of Animals for Slaughter*. Council of Europe, Strasbourg.
COE (continuing). *Information Bulletin on Legal Activities within the Council of Europe and in Member States*. Council of Europe, Strasbourg.
COE (1986). *Explanatory Report on the European Convention for the Protection of Vertebrate Animals used for Experimental and other Scientific Purposes*. Council of Europe, Strasbourg.
SCAW (continuing). *Newsletter*. Scientists' Center for Animal Welfare, Washington, DC.
WSPA (continuing). *Animals International*. World Society for the Protection of Animals, London.

Appendix 1 Useful Addresses

Agriculture Department of the Welsh Office, Park Avenue, Aberystwyth SY23 1NG
Animal Air Transport Association Inc., Box 55500, Fort Washington, MD 20744, USA
Animal Welfare Institute, P.O.Box 3650, Washington, DC 20007, USA
Association for the Study of Reptiles and Amphibians, Cotswold Wildlife Park, Burford, Glos OX8 4JW
Berne Convention Secretariat, Council of Europe, BP431R6, F-67006, Strasbourg Cedex, France
British Association for Shooting and Conservation, Marford Mill, Rossett, Wrexham, Clwyd LL12 OHL
British Field Sports Society, 59 Kennington Road, London SE1 7PZ
British Ornithologists' Union, c/o The Zoological Society of London, Regents Park, London NW1 4RY
British Small Animal Veterinary Association, 5 St. George's Terrace, Cheltenham, Glos GL50 3PT
British Laboratory Animal Veterinary Association, c/o British Veterinary Association (see below)
British Veterinary Association, 7 Mansfield Street, London W1M 0AT
British Veterinary Zoological Association, c/o British Veterinary Association (see above)
Cage and Aviary Birds, Specialist and Professional Press, 1 Throwley Way, Sutton, Surrey SM1 4QQ
Canadian Council on Animal Care, 1105–51 Slater Street, Ottawa, Ontario K1P 5H3, Canada
Canadian Wildlife Service, 9942 108 Street, Edmonton, Alberta, Canada
CITES Secretariat, 6 rue du Morpas, Case postale 78, CH-1000 Lausanne 9, Switzerland
Commission on Environmental Policy, Law and Administration, Adenauerallee 214, D5300 Bonn, Federal Republic of Germany
Convention on International Trade in Endangered Species Secretariat, 1196 Gland, Switzerland
Council of Europe, BP 431 R6, 67006 Strasbourg Cédex, France
Department of Agriculture and Fisheries for Scotland, Chesser House West, 500 Gorgie Road, Edinburgh EH11 3AW
Department of the Environment:
  RWA 3 Division, Room A504, Romney House, 43 Marsham Street, London SW1 P37
  Wildlife Conservation Licensing Section and International Trade in Endangered Species Branch, Tollgate House, Houlton Street, Bristol BS2 9DJ
Direction des Journaux Officiels, 26 rue Desaix, 75727 Paris Cédex 15, France

Environmental Law Centre, Adenauerallee 214, D5300 Bonn, Federal Republic of Germany
Eurogroup for Animal Welfare, 38 rue Georges Moreau, 1070 Brussels, Belgium
Expedition Advisory Centre, Royal Geographical Society, 1 Kensington Gore, London SW7 2AR
Farm Animal Welfare Council, Government Buildings, Hook Rise South, Tolworth, Surbiton, Surrey KT6 7NF
Fauna and Flora Preservation Society, c/o Zoological Society of London, Regents Park, London NW1 4RY
Fundação Brasileira para a conservaçâo de Natureza, Rua Miranda Valverde 103, Botafogo 22.281, Rio de Janeiro, Brazil
Health and Safety Executive Secretariat, General Enquiry Point, Regina House, Old Marylebone Road, London NW1 5RA
Her Majesty's Stationery Office, Publications Centre, P.O. Box 276, London SW8 5DT
Bookshops: 49 High Holborn, London WC1V 6HB; and at Edinburgh, Belfast, Manchester, Birmingham and Bristol
Home Office, Queen Anne's Gate, London SW1H 9AT
Humane Society of the United States, 2001 L St. NW, Washington DC 20057, USA
Institut Juridique International pour la Protection des Animaux, 86 rue du Pas Saint Georges, 33000 Bordeaux, France
Institute of Animal Technology, 5 South Parade, Summertown, Oxford OX2 7JL
Institute of Biology, 20 Queensberry Place, London SW7 2DZ
International Air Transport Authority, 26 Chemin de Joinville, P.O. Box 160, 1216 Cointrin, Geneva, Switzerland; 2000 Peel Street, Montreal, Quebec, Canada H3A 2R4
International Council for Bird Preservation, 219c Huntingdon Road, Cambridge CB3 0DL
International Union for Conservation of Nature and Natural Resources, 1196 Gland, Switzerland
Joint Advisory Committee on Pets in Society, Walter House, 418–422 Strand, London WC2R OPL
Mauritius Society for the Prevention of Cruelty to Animals, Moka Road, Rose Hill, Mauritius, Indian Ocean
Ministry of Agriculture, Fisheries and Food, Animal Health Division, Government Buildings, Hook Rise South, Tolworth, Surbiton, Surrey KT6 7NF
National Council for Aviculture, 87 Winn Road, Lee, London SE12 9EY
Nature Conservancy Council, Northminster House, Peterborough PE1 1UA
Nederlands Vereniging tot Bescherming van Vogels, Druebergseweg 160, 3708JB Zeist, Holland
Ramsar Secretariat, International Union for Conservation of Nature and Natural Resources, 1196 Gland, Switzerland
Research Defence Society, Lettsom House, 11 Chandos Street, Cavendish Square, London W1M 9DE
Royal College of Veterinary Surgeons, 32 Belgrave Square, London SW1X 8QP
Royal Society for the Prevention of Cruelty to Animals, Causeway, Horsham, Sussex RH12 1HG
Royal Society for the Protection of Birds, The Lodge, Sandy, Beds SG19 2DL
Scientists' Center for Animal Welfare, P.O. Box 9581, Washington, DC 20016, USA

Scottish Home and Health Department, New St Andrew's House, Edinburgh EH1 3YF

Secretariat of the World Heritage Committee, UNESCO, 7 place de Fonenoy, F-75700 Paris, France

Superintendent of Documents, US Government Printing Office (Supplier of Federal legislation), Washington, DC 20402, USA

UNEP/CMS Secretariat: Bonn Convention, Ahrstrasse 45, D-5300 Bonn 2, Federal Republic of Germany

United States Department of Agriculture, Washington, DC 20250, USA

United States Fish and Wildlife Service, US Department of the Interior, 18th and C Streets NW, Washington DC 20240, USA

Universities Federation for Animal Welfare, 8 Hamilton Close, South Mimms, Potters Bar, Herts EN6 3QD

Wildfowlers Association of Great Britain and Northern Ireland: now British Association for Shooting and Conservation (q.v.)

Wildlife Trade Monitoring Unit, IUCN Conservation Monitoring Centre, 219c Huntingdon Road, Cambridge CB3 ODL

World Society for the Protection of Animals, 106 Jermyn Street, London SW1Y 6EE; P.O. Box 190, 29 Perkins Street, Boston, MA 02130, USA

World Wildlife Fund, 1196 Gland, Switzerland

World Wildlife Fund — UK, 11-13 Ockford Road, Godalming, Surrey GU7 1QU

World Wildlife Fund — USA, 1601 Connecticut NW, Suite 800, Washington DC 20009, USA

Other addresses may be found in the "Useful Addresses" sections of

Deer Liaison Committee (1984). *Guidelines on the Safe and Humane Handling of Live Deer in Great Britain*. Deer Liaison Committee/Nature Conservancy Council, Peterborough

Cooper, J.E. and Hutchison, M.F. (eds) (1985). *Manual of Exotic Pets*. British Small Animal Veterinary Association, Cheltenham

Appendix 2 Scientific Names

Subspecies are excluded. In some cases a family or order is given. In a few instances an example is used, even though more than one species is involved. The common names follow those used in legislation and are not necessarily scientifically accurate.

MAMMALS

ALPACA	*Lama pacos*
ASS	*Equus asinus*
BADGER	*Meles meles*
BAT	Rhinolophidae/Vespertilionidae
BEAR, TEDDY	*Brunus edwardii*
BURRO	*Equus asinus*
CAMEL, ARABIAN	*Camelus dromedarius*
CAMEL, BACTRIAN	*Camelus bactrianus*
CAT, DOMESTIC, FERAL	*Felis catus*
CAT, SCOTTISH WILD	*Felis sylvestris*
CATTLE	*Bos taurus/Bos indicus*
COYPU	*Myocastor coypus*
DEER, FALLOW	*Dama dama*
DEER, MUNTJAC	*Muntiacus reevesi*
DEER, RED	*Cervus elaphus*
DEER, ROE	*Capreolus capreolus*
DEER, SIKA	*Cervus nippon*
DOG	*Canis familiaris*
DOLPHIN, COMMON	*Delphinus delphis*
DONKEY	*Equus asinus*
DORMOUSE	Gliridae
ELEPHANT	*Loxodonta africana/Elephas maximus*
FOX	*Vulpes vulpes*
GOAT	*Capra hircus*
GUINEA PIG	*Cavia porcellus*
HAMSTER, GOLDEN	*Mesocricetus auratus*
HARE	*Lepus capensis*
HEDGEHOG	*Erinaceus europaeus*
HINNY (hybrid of female ass and male horse)	

HORSE	*Equus caballus*
LLAMA	*Lama glama*
MINK	*Mustela vison*
MOLE	*Talpa europaea*
MOUSE	*Mus musculus*
MULE (hybrid of male ass and female horse)	
MUSK RAT	*Ondatra zibethicus*
OTTER, COMMON	*Lutra lutra*
PINE MARTEN	*Martes martes*
PIG	*Sus scrofa*
PORPOISE, HARBOUR	*Phocaena phocaena*
RABBIT	*Oryctolagus cuniculus*
SEAL, COMMON	*Phoca vitulina*
SEAL, GREY	*Halichoerus grypus*
SHEEP	*Ovis aries*
SHREW	Soricidae
SQUIRREL, RED	*Sciurus vulgaris*
SQUIRREL, GREY	*Sciurus carolinensis*
VICUNA	*Lama vicugna*
WEASEL	*Mustela nivalis*
WALRUS	*Odobenus rosmarus*
WHALE	Cetacea
WOLF	*Canis lupus*
ZEBRA	*Equus* sp.

BIRDS

BLACKBIRD	*Turdus merula*
BUDGERIGAR	*Melopsittacus undulatus*
CAIQUE	*Pionites* sp.
CAPERCAILLIE	*Tetrao urogallus*
COCKATIEL	*Nymphicus hollandicus*
COCKATOO	Psittacidae
CONURE	*Aratinga* sp.
COOT	*Fulica atra*
CROSSBILL	*Loxia curvirostra*
CROW	*Corvus corone*
DUCK, TUFTED	*Aythya fuligula*
EAGLE, BALD	*Aquila leucocephala*
EAGLE, GOLDEN	*Aquila chrysaetos*
FALCON, GYR	*Falco rusticolus*
FIELDFARE	*Turdus pilaris*
FIRECREST	*Regulus ignicapillus*
GADWALL	*Anas strapera*
GOLDENEYE	*Bucephala clangula*
GOLDFINCH	*Carduelis carduelis*
GOOSE, CANADA	*Branta canadensis*
GOOSE, GREYLAG	*Anser anser*
GOOSE, PINK-FOOTED	*Anser brachyrhynchus*
GOOSE, WHITE-FRONTED	*Anser albifrons*

GOSHAWK	*Accipiter gentilis*
GROUSE, BLACK	*Tegrao* or *Lyrurus tetrix*
GROUSE, RED	*Lagopus scoticus*
GUINEA FOWL	*Numida meleagris*
HAWK, HARRIS'S	*Parabuteo unicinctus*
KESTREL	*Falco tinnunculus*
LORY	Psittacidae
LORIKEET	Psittacidae
LOVEBIRD	*Agapornis* sp.
MACAW	*Ara* sp.
MALLARD	*Anas platyrhynchos*
MERLIN	*Falco columbarius*
MOORHEN	*Gallinula chloropus*
OWL, BARN	*Tyto alba*
PARROT	Psittaciformes
PARTRIDGE, GREY	*Perdix perdix*
PEACOCK (PEAFOWL)	*Pavo cristatus/muticus*
PEREGRINE	*Falco peregrinus*
PHEASANT	*Phasianus colchicus*
PIGEON, FERAL	*Columba livia*
PINTAIL	*Anas acuta*
PLOVER, GOLDEN	*Pluvialis apricaria*
POCHARD	*Aythya ferina*
PTARMIGAN	*Lagopus mutus*
QUAIL	*Coturnix coturnix*
REDWING	*Turdus iliacus*
SHOVELER	*Anas clypeata*
SNIPE	*Gallinago gallinago*
SPARROW, HOUSE	*Passer domesticus*
SPOONBILL	*Platalea leucorodia*
STARLING	*Sturnus vulgaris*
SWAN, MUTE	*Cygnus olor*
TEAL	*Anas crecca*
TURKEY	*Meleagris gallopavo*
WIGEON	*Anas penelope*
WOODCOCK	*Scolopax rusticola*
WOODPIGEON	*Columba palumbus*
YELLOWHAMMER	*Emberiza citrinella*

FISH

CARP	*Cyprinus carpio*
EEL	*Anguilla anguilla*
LAMPERN	*Lampetra fluviatilis*
SALMON	*Salmo salar*
TROUT	*Salmo trutta*
TROUT, RAINBOW	*Salmo irideus*

REPTILES AND AMPHIBIANS

ADDER	*Vipera berus*

FROG, COMMON	*Rana temporaria*
SNAKE, GRASS	*Natrix natrix*
SNAKE, RUBIK'S	*Naja rubika*
TOAD, AFRICAN CLAWED	*Xenopus* sp.
TOAD, COMMON	*Bufo bufo*

CRUSTACEANS

CRAYFISH, AMERICAN SIGNAL	*Pacifastacus leniusculus*

INSECTS

BEE, HONEY	*Apis mellifera*
BEETLE, COLORADO	*Leptinotarsa decemlineata*

Appendix 3 Recent Changes in Law and Addenda

NOTES

1 Specific terminology has been adopted by IUCN for the classification of the degree to which the threat of extinction applies to a species. Thus, all species at risk are referred to as "threatened". Within this generic term IUCN recognises and defines a number of status categories which it uses in its Red Data Books, a Domesday series for wildlife, e.g. Collins and Morris (1985), and which is followed elsewhere, including CITES (see Chapter 7). The terms are, in descending order of scarcity: extinct, endangered, vulnerable, rare, indeterminate, insufficiently known and out of danger.

2 *Legal status of animals*

Animals do not have legal personality and consequently cannot acquire or enforce, or have enforced on their behalf, any rights in law. The origins of this concept can be traced to principles of Roman law (Hume, 1957).

Conversely, animals do not have duties nor can they be accused of offences. In former times mediaeval church law in continental European countries recognised agreements between people and animals. Their ecclesiastical courts also tried animals, of many species, which were alleged to have caused damage to humans or their property (Hill, 1987; Evans, 1987).

Philosophical arguments have been advanced by Singer (1975), Griffiths (1982) and others in favour of animals having rights. The legal implications have been considered by Burr (1975, see Chapter 9), Morris and Fox (1978) and Favre and Loring (1983, see Chapter 9). Salt (1980) and Magel (1981) have produced bibliographies on the subject of animal rights.

Since animals have no legal status, rights and responsibilities with regard to them must be vested in their owners or keepers. Legislation for their benefit, such as welfare or conservation, can only impose requirements on, and enforce the law against, human beings.

The long-standing Scottish law on liability for injury and damage caused by dangerous animals was replaced by the Animals (Scotland) Act 1987 which brings Scottish law into line with that of England and Wales.

3 For the laws relating to Cetacea in general see Klinowska and Brown (1986).

4 *Animals on the highway*

The Animals Act 1971 provides for liability arising from damage caused by animals which are on the highway.

Liability for any animal which strays on to the highway is based on the general law of negligence (s. 8(1)). However, liability does not arise in the case of animals, which being lawfully kept on unfenced common land, stray from unfenced land on to a highway (s. 8(2)).

Although there is normally liability for damage caused by livestock which trespass on another person's land, nevertheless if such livestock stray during the lawful use of a highway, no liability arises (s. 5(4)). In the case of Matthews *v.* Wicks (1987) it was held that sheep, which wandered at will in a town where unfenced grazing was permitted and then entered an enclosed garden from the highway and caused damage, were not in lawful use of the highway when they wandered and the owner of the sheep was liable for the damage caused.

5 The long-standing Scottish law on liability for injury and damage caused by dangerous animals was replaced by the Animals (Scotland) Act 1987 which brings Scottish law into line with that of England and Wales.

6 The Highway Code (HMSO, 1987) provides advice for avoiding accidents. Dogs on the road should be kept on a lead and when in the car they must be under control. Horses and herded animals should be kept on the left on roads and after sunset lights and reflective clothing ought to be used. There are only limited rights for horses to be ridden on footpaths (see Clayden and Trevelyan, 1983).

7 The law regarding the conditions for merchantable quality and fitness for purpose has been reviewed by the Law Commission and Scottish Law Commission. Their Report (Law Commissions, 1987) has made recommendations for amendment of the relevant part of the Sale of Goods Act 1979.

8 There has been no judicial decision on the status of an animal which has been captured and held under temporary restraint (e.g. by physical means or chemical immobilisation) for field studies or first aid or veterinary treatment. However, the judgements in the cases involving accidental restraint state that capture must be accompanied by "acts of dominion" to fall within the definition of "captive animal" and hence within the Protection of Animals Acts. Common sense suggests that deliberate restraint and subsequent procedures would bring the animal within the Acts even though it is not intended to be permanent captivity.

If the detention of the animal is part of a duly authorised regulated procedure under the Animals (Scientific Procedures) Act 1986 the Acts will not apply (see Chapter 4, p. 77).

9 The Protection of Animals (Penalties) Act 1987 raised the fines which can be imposed in respect of cruelty offences to imprisonment for up to six months and/or a fine not exceeding level 5 on the standard scale.

In 1987 there was an abortive attempt to amend certain aspects of the Protection of Animals Act 1911. A Private Member's Bill provided for disqualification from keeping any kind of animal on a first conviction for cruelty; for police powers of entry to see if an offence has been committed; and for a new definition of "captive" (see p. 28 and Appendix 3 earlier) to include wild animals which are incapable of escaping (RSPCA, 1987).

10 The Welfare of Livestock (Prohibited Operations) (Amendment) Regulations 1987 No. 114 make tooth grinding in sheep a prohibited operation.

11 A veterinary inspector has power under the Order to issue a certificate permitting the animal's movement.

12 Greenwood (1987) has discussed the effectiveness of the Act, particularly with respect to inspections and Collins (1987, in Chapter 3) has described the application of the Act to butterfly houses. Klinowska (1987) and Klinowska and brown (1986) have reviewed the legislation affecting Cetacea, including the Zoo Licensing Act.

The Local Government (Access to Information) Act 1985 gives public access to documents presented at certain local authority council and committee meetings. Such documents may include reports made under the foregoing Acts.

13 Animals (Scientific Procedures) Commencement Order 1986 brought all parts of the Act into force except ss. 7, 10(3) and Schedule 2 (relating to designated breeding and supplying establishments and s. 29 (Northern Ireland).

14 The Animals (Scientific Procedures) Act (Fees) Order 1986 set these fees at £100 plus £67 for each personal licence as from 1 January 1987.

15 Other identification methods which can interfere with sensitive tissues such as ear clipping, some fin tagging and branding are also regulated procedures.

16 The Abandonment of Animals Act 1960 must be considered (see Chapter 3).

17 Collins (1987) has reproduced the MAFF "free list" of insects which may be imported and sold without a licence. The list is liable to alteration.

18 Any animal which is taken into captivity becomes subject to the Protection of Animals Acts 1911–1964 (see Chapter 3). The question of cruelty to wild animals and the implications of capture and temporary restraint have been examined in Chapter 3, p. 28, Chapter 4, p. 72 and in the notes thereto in this Appendix.

19 Consideration is being given to extending protection under the Wildlife and Countryside Act to the wild cat, dormouse, pine marten, walrus, all Cetacea and certain reptiles. The first three and the harbour porpoise are already Schedule 6 species which may not be taken by certain methods. Other Schedule 6 species include the badger, all bats, hedgehog, otter, polecat, all shrews and red squirrel. See also BFSS (1986, in Chapter 7).

20 The Crossbows Act 1987 imposes restrictions on the possession or acquisition of crossbows by people under 17 years old.

21 Derivatives of species which are subject to import licence include any part of, or anything made from, such species which can in any way be identified as such.

22 Anon (1987) provides a survey of European national legislation on insect conservation.

23 The Standing Committee of the farming animal Convention has made Recommendations on domestic fowls and pigs and is preparing one on cattle (COE, 1987).

24 The complaint of the RSPCA was upheld by the EEC Commission which then referred the UK to the European Court for its failure to enforce, and in the case of France, for its failure to implement, the EEC Directives on transportation (EEC, 1977, 1981).

Farm animal welfare has recently been reviewed by the EEC in a Public Hearing in 1986 and in a Report (Simmonds, 1986) culminating in a Resolution of the Parliament (EEC, 1986).

25 *Belgium*

New animal welfare legislation replaces the Act of 1972 as from 1 December 1987.

Federal Republic of Germany

The Animal Protection Act 1986 which came into force on 1 January 1987 makes provision for the welfare of domestic and captive animals.

26 WSPA (1981) surveyed the legal provisions in western Europe for the use of animals in making films.

27 *The Netherlands*

The Netherlands' Experiments on Animals Act was brought into effect by a Decree of 31 May 1985 and the two provisions implement the European legislation. See also Walvoort (1986).

Switzerland

The Swiss law is discussed by Goetschel (1986, see Chapter 9) and Weber (1986) has examined the implications of the 1985 Referendum for the abolition of animal experimentation (which failed) and other legal and ethical issues.

28 *Federal Republic of Germany*

The 1972 Act was replaced by the Animal Protection Act 1986 which came into effect on 1 January 1987. This brings West German law into line with the European legislation.

29 *Belgium*

New animal welfare legislation replaces the Act of 1972 as from 1 December 1987.

30 See IUCN (1983; 1986) and Cooper (in preparation).

Information on the Conventions can be obtained from their Secretariats. Meetings of the states members are held, usually at five-yearly intervals, of which reports are published (see, for example, CMS, 1985; CITES, various).

31 See also IUCN (1985; 1986) and Osterwoldt (1986).

Conferences on the major conservation Conventions are held from time to time, e.g. CITES (various), CMS (1985). Information on their progress and amendments can be obtained from their Secretariats.

32 See also Anon (1987).

Some governments publish translations of their national legislation as do, for example, Germany (Ellers, 1985) and Norway (Anon, 1985). For a review of Russian wildlife legislation, see Pryde (1986).

33 Contemporary papers, published after 1980, include studies of African and Scandinavian laws, and conservation conventions.

CASE

Matthews *v*. Wicks (1987). *The Times*, 25 May 1987.

REFERENCES

Anon (1985). *The Wildlife Act*. Ministry of Environment, Oslo.

Anon (1987). *Legislation to Conserve Insects in Europe*. Amateur Entomologists' Society, London.

CITES (various). Proceedings of meetings and extraordinary meetings. CITES Secretariat, Lausanne.

Clayden, P. and Trevelyan, J. (1983). *Rights of Way: A Guide to Law and Practice*. Open Spaces Society and Ramblers' Association, London.

CMS (1985). *Convention on the Conservation of Migratory Species of Wild Animals: CMS. Proceedings of the First Meeting of the Conference of the Parties*. Secretariat of the Convention, Bonn.

COE (1987). *Information Bulletin on Legal Activities within the Council of Europe and in Member States* 24, 22–23.

Collins, N.M. (1987). MAFF/DAFS Free List. Annex 3. In *Butterfly Houses in Britain. The Conservation Implications*. International Union for Conservation of Nature and Natural Resources, Gland and Cambridge.

Collins, N.M. and Morris, M.G. (1985). *Threatened Swallowtail Butterflies of the World. The IUCN Red Data Book*. International Union for Conservation of Nature and Natural Resources, Gland and Cambridge.

Cooper, M.E. (in preparation). The law relating to migratory birds. In *Proceedings of the III World Conference* (Chancellor, R.C. and Meyburg, B.-U., eds). World Working Group on Birds of Prey and Owls of the International Council for Bird Preservation, London, Paris, Berlin.

EEC (1977). Council Directive 77/489/EEC of 18 July 1977 on the protection of animals during international transport. *Official Journal of the European Communities* of 8 August 1977 No. L 200, 10–16. Council of Ministers of the European Communities, Brussels.

EEC (1981). Council Directive 81/389/EEC of 12 May 1981 establishing measures for the implementation of Directive 77/489/EEC on the protection of animals during international transport. *Official Journal of the European Communities* of 6 June 1981, No. L 150, 1–5. Council of Ministers of the European Communities, Brussels.

EEC (1986). European Parliament Resolution on animal welfare policy. Document A-211/86. *Official Journal of the European Communities* of 23 March 1987 No. C76/186. European Economic Communities, Brussels.

Ellers, W. (ed.) (1985). *Federal Hunting Act*. Federal Ministry of Food, Agriculture and Forestry, Bonn.

Evans, E.P. (1987). *Criminal Prosecution and Capital Punishment of Animals*. Faber, London.

Greenwood, A.G. (1987). Zoo inspections — an effective welfare measure? In *The Welfare of Animals in Captivity*. British Veterinary Association, London.

Griffiths, R. (1982). *The Human Use of Animals*. Grove, Bramcote.

Hill, R. (1987). *Both Small and Great Beasts*, 2nd edn. Universities Federation for Animal Welfare, Potters Bar.

HMSO (1987). *The Highway Code*, revised edn. HMSO, London.
Hume, C.W. (1957). *The Status of Animals in the Christian Religion*. Universities Federation for Animal Welfare, London, now Potters Bar.
IUCN (1983). *Elements of an Agreement on the Conservation of Western Palearctic Migratory Species of Wild Animals*. Environmental Law and Policy Paper 21. International Union for Conservation of Nature and Natural Resources, Gland and Cambridge.
IUCN (1985). *Status of Multilateral Treaties in the Field of Environment and Conservation*. Environmental Law and Policy Occasional Paper 1. International Union for Conservation of Nature and Natural Resources, Gland and Cambridge.
IUCN (1986). *Migratory Species in International Instruments*. Environmental Law and Policy Occasional Paper 2. International Union for Conservation of Nature and Natural Resources, Gland and Cambridge.
Klinowska, M. (1987). Cetaceans in captivity. In *The Welfare of Animals in Captivity*. British Veterinary Association, London.
Klinowska, M. and Brown, S. (1986). *A Review of Dolphinaria*. Department of the Environment, Bristol.
Law Commissions (1987). *Sale and Supply of Goods*. Cmnd 137. HMSO, London.
Magel, C.R. (1981). *A Bibliography on Animal Rights and Related Matters*. University Press of America, Washington DC.
Morris, R.K. and Fox, M.W. (1978). *On the Fifth Day*. Acropolis, Washington DC.
Osterwoldt, R. (1986). International agreements. *Naturopa* **54**, 11–13.
Pryde, P.R. (1986). Strategies and problems of wildlife preservation in the USSR. *Biological Conservation* **36**, 351–374.
RSPCA (1987). Reform of the Protection of Animals Act, 1911. *RSPCA Today* **55**, 20–21.
Salt, H.S. (1980). *Animal Rights*. Society for Animal Rights, Clarks Summit, Pa.
Schoenbaum, M. and Benhar, E. (eds) (1985). *Ethical Aspects of the Use of Animals in Research*. Special issue. *Quarterly of the Israel Zootechnical Association* **14**, 1–2.
Simmonds, R. (1986). Report on animal welfare policy. *European Parliament Working Documents 1986–87* 19 January 1987 Series A, Document A 2–211/86. European Economic Communities, Brussels.
Singer, P. (1975). *Animal Liberation*. Random House, New York.
Walvoort, H.C. (1986). Contribution of pathology to laboratory animal welfare. *Laboratory Animals* **20**, 291–292.
Weber, H. (1986). Democratic expression of public opinion on animal experimentation. *Journal of Medical Primatology* **15**, 379–389.
WSPA (1981). Rating films from an animal welfare point of view. *Animals International* **1**(4), 8–9.

Index

For legislation see Table of Statutes, Table of Statutory Instruments and Table of Other Legislation. Not all species of animal are listed. Reference should be made to Appendix 2 which gives English and scientific names.